AutoCAD Certified User Study Guide

William Wyatt
Daniel John Stine

SDC Publications
P.O. Box 1334
Mission, KS 66222
913-262-2664
www.SDCpublications.com
Publisher: Stephen Schroff

ISBN-13: 978-1-63057-598-4
ISBN-10: 1-63057-598-4

Printed and bound in the United States of America.

Foreword

The intent of this book is to provide the reader with a **study guide** and electronic **practice exam** (download) to prepare for the Autodesk® AutoCAD® Certified User Exam. You will find an overview of the exam process, the user interface, and the official exam objectives: Draw and Modify Objects, Draw with Accuracy, Basic Editing, Annotation, and Layouts & Printing.

While this study guide cannot claim to cover every possible question that may arise in the exam, it does help to firm up your basic knowledge to deal with most questions positively… thus, providing more time to reflect on the more difficult questions.

> **Errata:**
> Please check the publisher's website from time to time for any errors or typos found in this book after it went to the printer. Simply browse to www.SDCpublications.com, and then navigate to the page for this book. Click the **View/Submit errata** link in the upper right corner of the page. If you find an error, please submit it so we can correct it in the next edition.
>
> You may contact the publisher with comments or suggestions at service@SDCpublications.com.

You may contact the publisher with questions, comments, or suggestions at service@SDCpublications.com.

Exercise Files

The drawing files required for the tutorials may be downloaded from SDC Publication's website following the instructions on the inside front cover of this book.

Trial and Student Software:

This book is based on Autodesk *AutoCAD 2024*. A **20-day** trial may be downloaded from Autodesk's website. Additionally, qualifying students may download the free **1-year** student version of the software from **students.autodesk.com**. Both are fully functional versions of the software. The provided practice exam, as well as the official exam, require the use of AutoCAD and the provided AutoCAD drawing files to answer most questions successfully.

About the Authors

William G. Wyatt, Sr. is an instructor and Program Head for the Architectural Engineering Technology program at John Tyler Community College in Chester, Virginia. He earned his Doctor of Education degree from Virginia Tech and his Master of Science and Bachelor of Science degrees in industrial technology from Eastern Kentucky University. He earned his associate degree in architectural technology from John Tyler Community College. He is a certified Architectural and Building Construction Technician, AutoCAD Certified User and Revit Certified User. He is the author of twelve texts of Autodesk AutoCAD Architecture and AutoCAD Certification.

Daniel John Stine is a registered Architect (WI) and Director of Design Technology at the top-ranked architecture firm Lake|Flato in San Antonio, Texas. Dan is a member of the American Institute of Architects (AIA), Construction Specification Institute (CSI) and has previously taught AutoCAD and Revit classes at Lake Superior College. He currently teaches Revit to graduate Architecture students at North Dakota State University (NDSU); additionally, he is a Certified Construction Document Technician (CDT). He has presented at Autodesk University in Las Vegas and internationally via the BILT Conference.

Table of Contents

Chapter 2: Drawing Objects

Chapter 3: Drawing with Accuracy

Chapter 4: Organizing Objects

Chapter 6: Using Additional Drawing Techniques

Chapter 7: Annotating Drawings

Chapter 8: Reusing Existing Content

Notes:

Chapter 1
Applying Basic Drawing Skills

This chapter will review the User Interface of AutoCAD 2024 and basic drawing setup, including setting the drawing units, drawing limits, and an overview of paper and model spaces in preparation for exam questions. Following the review material, there is a tutorial and two quizzes that will help you prepare for the test.

The student will be able to:
1. Start a new drawing from a template
2. Identify the current workspace
3. View and set the Units of the drawing using the Status bar
4. Set the Drawing Limits
5. Identify the major components of the Ribbon
6. Turn on specified toggles of the Status bar
7. Modify the display of the ribbon panels to Tabs, Panel Titles, and Panel Buttons
8. Draw lines in model space and view the design to scale in a Layout
9. Edit the Page Setup Manager to print to a pdf file of a Layout
10. Set the Grid and Snap settings for a drawing
11. Identify the following User Interface components: Ribbon, Quick Access toolbar, Drawing Tabs, Viewport Control, ViewCube, Navigation Bar, Command Line, UCS Icon, and Status bar
12. Create a selection set of entities
13. Use the Measure tools of the Inquiry panel in the Home tab

Components of the AutoCAD 2024 Interface

When you launch AutoCAD 2024, it opens to the Start tab, as shown in Figure 1-1.

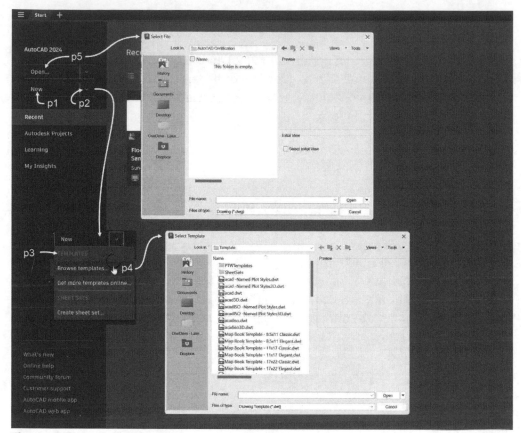

Figure 1-1

To start a new drawing, choose the New button shown at p1 as shown in Figure 1-1 to start a new drawing from the default acad.dwt template. You can choose the chevron shown at p2 to expand the New button menu, which allows you to choose the acad.dwt template from the drop-down menu shown at location p3 in Figure 1-1 or select the Browse templates... option to display the list of templates shown at location p4 in Figure 1-1.

To Open an existing drawing choose the Open button shown at p5 in Figure 1-1 to access the Select file dialog box.

The Learning button shown at p1 in Figure 1-2 opens to Online Help and Tips and Video materials as shown in Figure 1-2.

Figure 1-2

Overview of the User Interface

The test typically includes questions to identify the names of major components of the workspace, as shown in Figure 1-3.

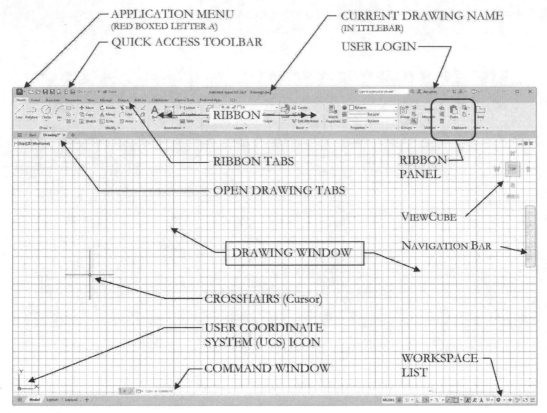

Figure 1-3

Ribbon

The ribbon is at the top of the workspace; it consists of tabs to group a set of panels. Commands with a similar purpose are grouped into panels. The Home tab includes panels such as Draw, Modify, and Annotation panels. As shown in Figure 1-4, the Draw panel includes commands to create lines, circles, and arcs whereas the Modify panel includes commands to Move, Copy, and Rotate existing entities.

Figure 1-4

You can left-click on the panel title and drag it to the drawing window, as shown in Figure 1.5. To return the panel to its original position in the ribbon, hover the cursor over the panel to display the right margin and select the Return Panel to Ribbon button shown in Figure 1-5.

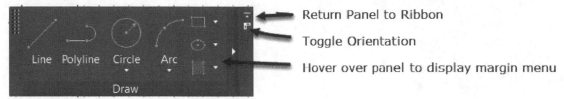

Figure 1-5

Quick Access Toolbar

Above the ribbon is the Quick Access toolbar, which includes commands such as QNew, Open, Save, SaveAs, Print, Undo, and Redo. You can add or delete commands from the menu drop down as shown at p1 in Figure 1-6. If you check the Workspace from the list of options, the current workspace will be shown in the Quick Access Toolbar as shown at p2 in Figure 1-6.

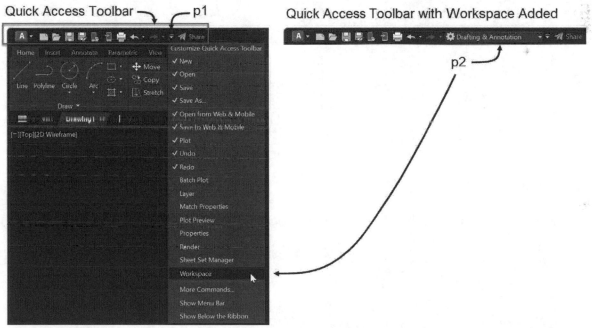

Figure 1-6

Application Menu

The Application Menu can be accessed by clicking the red **A** located in the upper left corner of the workspace as shown at p1 in Figure 1-7. The menu includes commands for saving, printing and exporting the current file. The command search window shown at p2 allows you to type a command to view where to access the command.

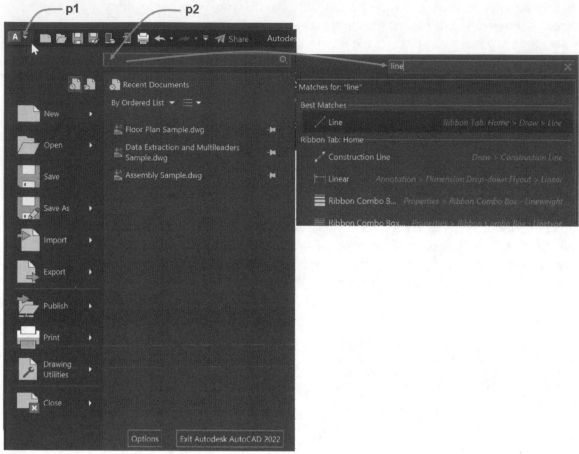

Figure 1-7

Accessing Workspaces

Named workspaces define the content of the Ribbon. The current workspace can be displayed in the Quick Access toolbar as shown in Figure 1-8 (if adjusted per Figure 1-6).

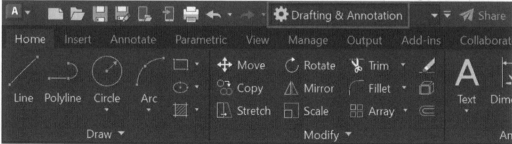

Figure 1-8

The flyout menu of the workspace window shown in Figure 1-9 lists the names of available saved workspaces, and the Customize... option opens the Customize User Interface dialog box, which is used to create new workspaces.

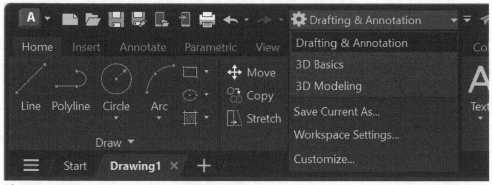

Figure 1-9

You may select workspaces by left-clicking the Workspace Switching toggle of the Status bar, as shown in Figure 1-10.

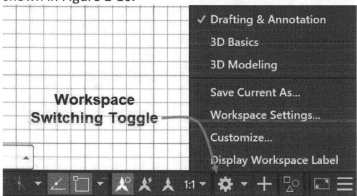

Figure 1-10

Changing Panel Display

You can toggle the ribbon display from Full Panel Display to Panel Titles, Panel Buttons, or Tabs by choosing the Ribbon Display toggle shown at the right of the last tab at p1 in Figure 1-11.

Figure 1-11

The display of the panel options is shown in Figure 1-12.

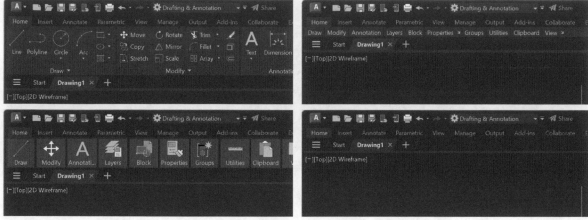

Figure 1-12

Accessing Help

You can access Help by choosing the "?" question mark located above the ribbon at right, as shown in Figure 1-13. You can type a command in the Info Center field to view topics in the knowledge base and help.

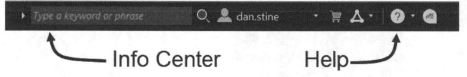

Figure 1-13

When the Help window opens you can type a keyword for help, as shown in Figure 1-14 at right, or select from the topic list shown at left.

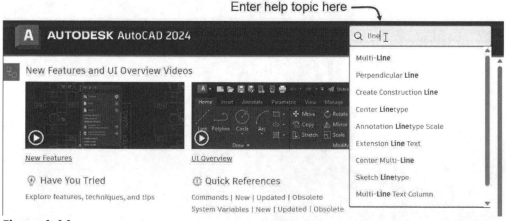

Figure 1-14

Drawing Tabs

The Drawing tabs are located immediately below the ribbon at left as shown in Figure 1-15. You can right-click over the tabs to display a contextual menu. The tabs allow you to switch to other open drawings quickly. The drawing tabs can be dragged to a second monitor, where the drawing may be edited outside of the primary application window. You can create a new drawing from the contextual menu or choose the plus sign "+" shown at p1 in Figure 1-15.

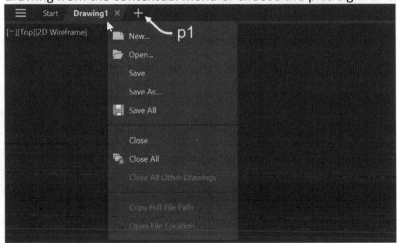

Figure 1-15

Viewport Control Menu

The Viewport Control menu is located below the ribbon at left as in Figure 1-16.

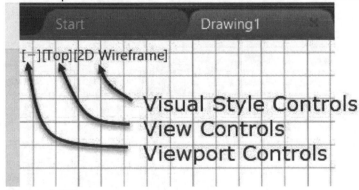

Figure 1-16

The Viewport Controls (-) menu shown in Figure 1-17 includes tools for the creation of model space viewports using the Viewport Configuration List and commands for turning on / off the display of ViewCube, Steering Wheels and the Navigation Bar.

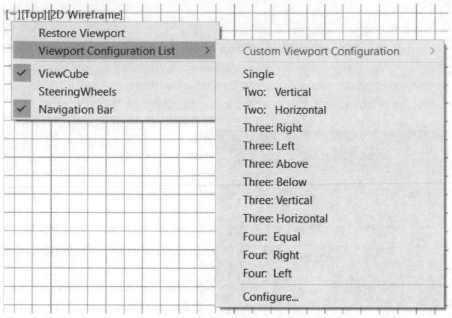

Figure 1-17

The View Controls menu includes tools similar to the tools of the ViewCube, as shown at left in Figure 1-18. The last menu is the Visual Styles, which provides quick access to the visual styles as shown at right in Figure 1-18. Simply left click on the label to access these menus.

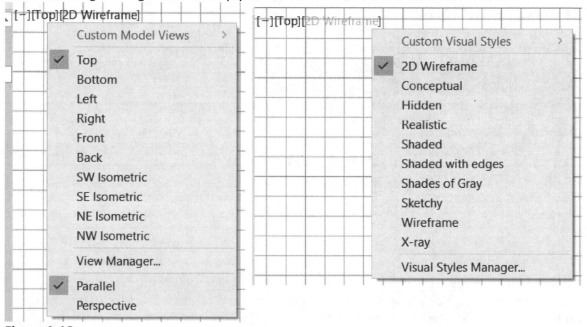

Figure 1-18

Using the ViewCube and Navigation Bar

The ViewCube and Navigation Bar are located immediately below the ribbon at right in the Drawing Window. An orthographic view, such as "front" can be obtained by selecting the S as shown in Figure 1-19. Choose the Clockwise or Counter-Clockwise icons (only visible when hovering the cursor over the ViewCube) to rotate the ViewCube and the resulting views.

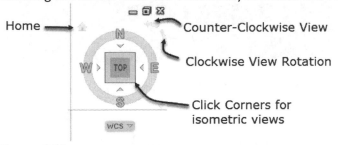

Figure 1-19

The Navigation Bar shown in Figure 1-20 provides quick access to viewing tools such as Pan and Zoom. The Zoom flyout includes options for the Zoom command, as shown in Figure 1-20.

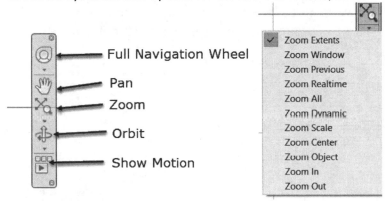

Figure 1-20

Command Line

The Command Line (located above the Status bar) shown in Figure 1-21 provides space to type commands and see what steps are required/next while in certain commands. The program responds to your command by listing prompts, options, and warnings regarding commands, as shown in Figure 1-21.

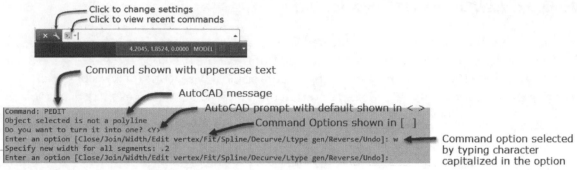

Figure 1-21

Model and Layout Tabs

The Model tab is the location that entities, such as lines, arcs and circles are created and annotation added to create the drawing full size. The Layout tab shown in Figure 1-22 is used for scaling and printing all or part of the drawing you created in the Model tab. A layout may include a titleblock and border for a sheet of paper. A layout includes one or more viewports that allow you to control the scale and what part of the model drawing that is displayed in a layout. Think of a viewport as a view from a layout (aka paper space) into model space.

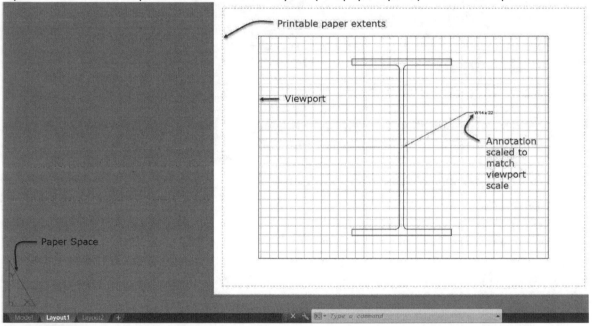

Figure 1-22

To activate floating model space while in a layout, double-click inside the viewport near p1 shown in Figure 1-23. The model space UCS will be displayed in the lower-left corner of the viewport, as shown at p2 in Figure 1-23, and the Viewport scale is displayed in the Status bar.

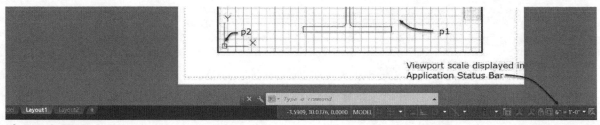

Figure 1-23

Status bar

The Status bar is located below the Drawing Window. The Status bar includes toggles, which allow you to turn on or off commands such as Grid, Ortho, and Object Snaps. The far-left field lists the model or paper space of your crosshair; the Model tab was current when the image was captured in Figure 1.24. The far-right field of the status bar is the Customization button; when clicked, it will allow you to select from a complete list of toggles for display in the Status bar.

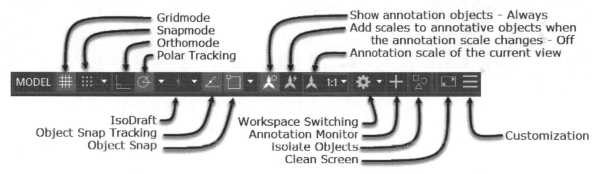

Figure 1-24

Drawing Setup

Prior to drawing anything in model space, you should do the following three things: set the drawing units, define the drawing limits to fit the item that you will create and then choose Zoom All to zoom the model space to match the drawing limits. It is possible to create a custom template drawing, which consists of settings for units, limits, layers, and styles for text and dimensions to reduce drawing setup if you are creating similar type drawings.

Setting Units

The units for the acad.dwt template is Decimal. You can display the units in the Status bar if you select the Customization button located on the far right of the status bar and choose Units as shown at p1 in Figure 1-25 (fyi: model tab is current for this screen capture).

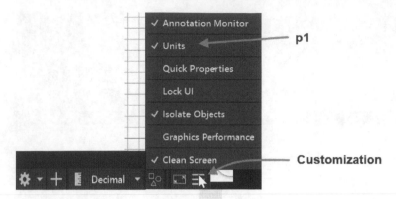

Figure 1-25

There are 5 types of units included in the flyout of the Units toggle of the Status bar as shown in Figure 1-26.

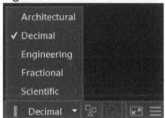

Figure 1-26

To view and edit the settings of a unit type click the Application Menu shown at p1 in the upper left corner of the graphics screen and choose Drawing Utilities > Units from the Application Menu shown in Figure 1-27.

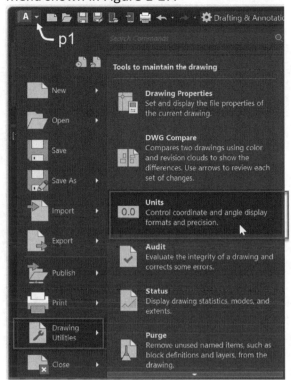

Figure 1-27

When the Drawing Units dialog box opens you can change the precision as shown in Figure 1-28 at p1 for lengths and at p2 for angle measures. Changes in the Drawing Units dialog box control the format of the information presented in the Properties palette, Coordinates fields of the Status bar and the Command Line.

Figure 1-28

Therefore, if the length precision decreased to two decimal places and the angle precision increased to one decimal place, the length and angle information when drawing a line will change in the dynamic input fields as shown in Figure 1-29. When you select the line, the length and angle information is listed as set in Drawing Units settings in the Properties palette, as shown in Figure 1-29.

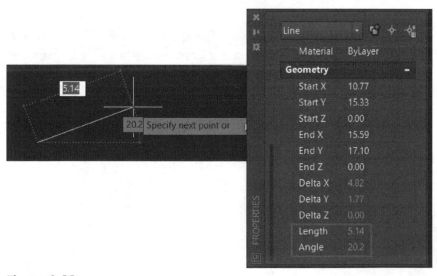

Figure 1-29

Setting Drawing Limits

The Limits command allows you to specify the extents of the drawing. The default limits are set in the acad.dwt template to 12" x 9". If designing a part that is 10 feet in diameter most of the part would not be visible using the acad.dwt template. Therefore, you should set the limits greater than the part size. To view the extents of limits in your graphics screen to match limits settings choose the Zoom All command.

Choose the Drawing Limits command by opening the Application Menu and typing **Lim** in Command Search Window shown in Figure 1-30. The Drawing Limits command will display below the Command Search Window, allowing you to access the command from the list as shown in Figure 1-30.

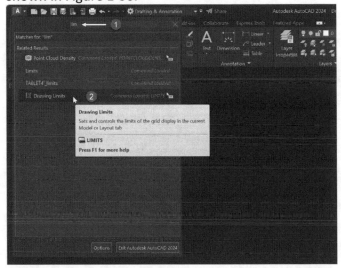

Figure 1-30

When you access the command you are prompted to specify the X,Y values for lower left corner and the upper right corner in the workspace prompts and the Command Window as follows:
Command: '_limits
Reset Model space limits:
Specify lower left corner or [ON/OFF] <0.00,0.00>: **0,0** *(Type **0,0** press Enter to specify the X,Y value for the lower left corner of the drawing window.)*
Specify upper right corner <12.00,9.00>: **150,150** *(Type **150,150** press Enter to specify the X,Y value for the upper right corner of the drawing window.)*

Viewing Cursor Locations with Coordinates

If you choose the Customization button of the Status bar shown in Figure 1-31, you can turn on the Coordinates to view the X,Y,Z values of your cursor as you move it in the drawing window.

If you move your cursor over the User Coordinate System Icon located in the lower left corner, the coordinates will be approximately 0,0,0 in the Coordinates field, as shown in Figure 1-32.

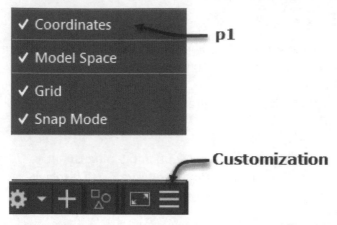

Figure 1-31

Figure 1-32

Using Zoom All to View the Limits of the Drawing

If you move the cursor over the upper right corner, the coordinates will not be 150,150 until you choose Zoom All. Zoom All will display all visible entities of the drawing and the Limits. Zoom All can be selected from the Navigation bar shown at the right border of the graphics screen as shown in Figure 1-33.

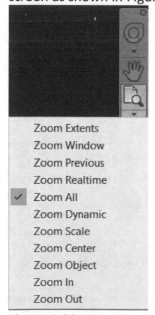

Figure 1-33

Creating a Selection Set

The pick box is included in the cursor of AutoCAD, as shown at p1 in Figure 1-34, which allows you to select any entity of the drawing, such as lines, circles, and rectangles. The Properties palette is displayed by choosing the Properties button at p2 in Figure 1-34 of the Palettes panel in the View tab. The Select command shown at p3 in Figure 1-34, which allows you to create a selection set for editing, is accessed from the Properties palette.

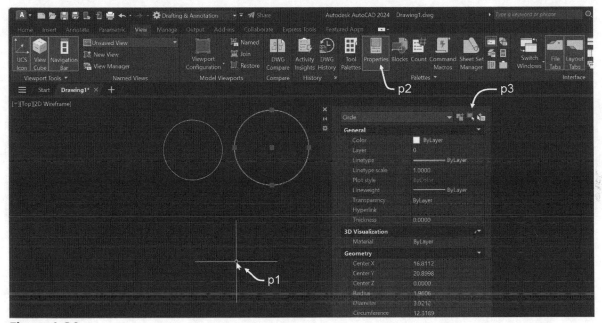

Figure 1-34

The Properties palette may also be displayed by choosing the launching arrow located on the Properties panel title from the Home tab, as shown at p1 in Figure 1-35, or typing Ctl+1 on the keyboard.

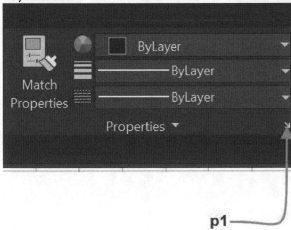

Figure 1-35

The Select command includes the following modes to create the selection set:
Window/Last/Crossing/BOX/ALL/Fence/WPolygon/CPolygon/Group/Add/Remove/Multiple/Previous/Undo/AUto/Single.

You can use the Add and Remove options to edit the objects included in the set. The Last mode adds the last object placed in the drawing whereas the Previous activates the previous selection set.

Note that you can open the **Select Demo** file from the **AutoCAD 2024 Certified User Exercise Files\ Exercise Files \ Certification Demo \ Ch 1** folder to try the technique to create a selection set as you read the following steps.

Choose the View tab. Choose the Properties Palette button of the Palettes panel as shown in Figure 1-36.

Figure 1-36

Choose the Select Objects button as shown in Figure 1-37 of the Properties palette.

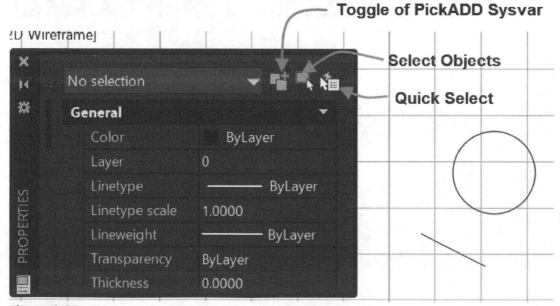

Figure 1-37

PSELECT

Select objects: L 1 found (Type **L** and press Enter to select the last entity placed in the drawing. The polygon shown at p1 in Figure 1-38 was the last entity created.)

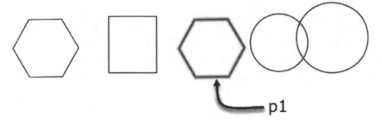

Figure 1-38

Select objects: 1 found, 2 total (Select the rectangle shown at p2 in Figure 1-39.)
Select objects: 1 found, 3 total (Select the polygon shown at p3 in Figure 1-39.)

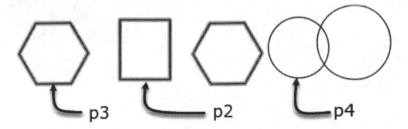

Figure 1-39

Select objects: **R** (Type **R** and press Enter to remove objects from the selection set, select the polygon shown at p3 in Figure 1-39.)
Remove objects: 1 found, 1 removed, 2 total
Remove objects: **A** (Type **A** and press Enter to add objects to the selection set, select the polygon shown at p4 in Figure 1-39.) Select objects: 1 found, 3 total
Select objects: (Press Enter to end selection.)
Command: (Press Enter to repeat the last command.)
PSELECT
Select objects: **P** (Type **P** and press Enter to choose the Previous selection set as shown in Figure 1-40.)

Figure 1-40

Additional material regarding selecting objects and selection sets is included in Chapter 3. The use of the Properties palette and Quick Select applications are included in Chapter 4.

Using Inquiry Tools

Inquiry tools are located on the Home tab in the Measure flyout of the Utilities panel as shown at left in Figure 1-41. The tools included within the MeasureGeom command allow you to measure distances and angles between entities. The test may use the tools to test the completion of moving entities by asking you to enter the new distance between reference points. The Quick tool shown in the flyout allows you to move the cursor to display distances as shown at right or select other measure tools. Tools are included for the area and volume calculations.

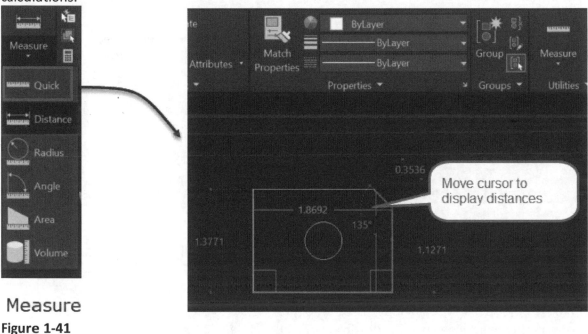

Measure

Figure 1-41

Distance and Area with Subtraction

The technique to use the Distance and Area of the MeasureGeom command including subtraction of multiple areas is included below. Note that you can open the **Inquiry demo** file from the **AutoCAD 2024 Certified User Exercise Files\ Exercise Files \ Certification Demo \ Ch 1** folder to follow the technique to determine distance and area as you read the following steps.

Click the flyout as shown at p1 in Figure 1-42, and verify the Endpoint object snap mode is toggled ON (it should have a checkmark next to it).

Verify Dynamic Input is toggled ON in the Status bar as shown at p1 in Figure 1-43. This toggle likely needs to be made visible first via the Customization menu to the far right.

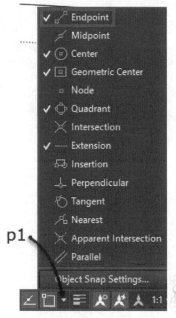

Figure 1-43

Figure 1-42

Choose the **Distance** tool shown in Figure 1-44 from the Measure flyout in the Utilities panel in the Home tab.

Figure 1-44

Respond to the workspace prompts as follows:

> Specify first point: (Move the cursor to p1 as shown in Figure 1-45 to display the green Endpoint Object Snap marker and click to specify the point.)
>
> Specify second point or: (Move the cursor to p2 as shown in Figure 1-45 to display the green Endpoint Object Snap marker and click to specify the point.)

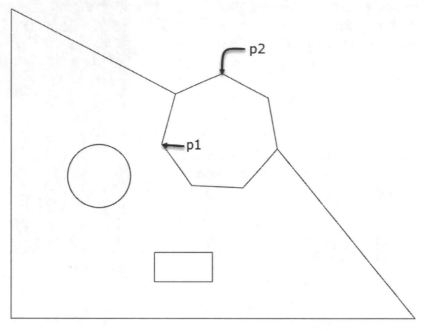

Figure 1-45

The results of the measure tool are displayed as 1.5789 in the workspace as shown in Figure 1-46, and in the command line.

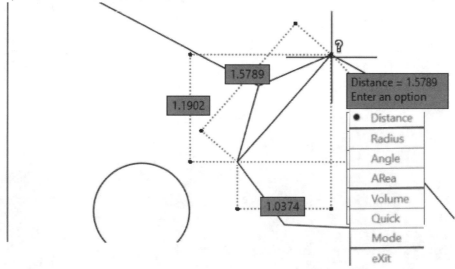

Figure 1-46

You can press Function key F2 to display the information in the Command Window as shown in Figure 1-47.

```
Command:
Command: _MEASUREGEOM
Enter an option [Distance/Radius/Angle/ARea/Volume] <Distance>: _distance
Specify first point:
Specify second point or [Multiple points]:
Distance = 1.5789,  Angle in XY Plane = 49,  Angle from XY Plane = 0
Delta X = 1.0374,  Delta Y = 1.1902,  Delta Z = 0.0000
Enter an option [Distance/Radius/Angle/ARea/Volume/Quick/Mode/eXit] <Distance>:
```
```
X ⚒ 🔲▼ MEASUREGEOM Specify first point: |
```

Figure 1-47

In the following steps, you will create the area of a shape, and subtract the areas of a circle and rectangle within the shape using Add and Subtract options with the Area tool of the MeasureGeom command. Choose the Area tool shown in Figure 1-48 from the Measure flyout in the Utilities panel in the Home tab.

Figure 1-48

Specify first corner point or [Object/Add area/Subtract area/eXit] <Object>: **A** (Type **A** to specify the ADD area option.)
Specify first corner point or [Object/Subtract area/eXit]: (Click to select the point at p1 in Figure 1-49 using the Endpoint Object Snap.)
 (ADD mode)Specify next point or [Arc/Length/Undo]: (Click to select the point at p2 in Figure 1-49 using the Endpoint Object Snap.)
 (ADD mode)Specify next point or [Arc/Length/Undo]: (Click to select the point at p3 in Figure 1-49 using the Endpoint Object Snap.)

(ADD mode)Specify next point or [Arc/Length/Undo/Total] <Total>: (Click to select the point at p4 in Figure 1-49 using the Endpoint Object Snap.)

(ADD mode)Specify next point or [Arc/Length/Undo/Total] <Total>: (Click to select the point at p5 in Figure 1-49 using the Endpoint Object Snap.)

(ADD mode)Specify next point or [Arc/Length/Undo/Total] <Total>: (Click to select the point at p6 in Figure 1-49 using the Endpoint Object Snap.)

(ADD mode)Specify next point or [Arc/Length/Undo/Total] <Total>: (Click to select the point at p7 in Figure 1-49 using the Endpoint Object Snap.)

(ADD mode)Specify next point or [Arc/Length/Undo/Total] <Total>: (Click to select the point at p8 in Figure 1-49 using the Endpoint Object Snap.)

(ADD mode)Specify next point or [Arc/Length/Undo/Total] <Total>: (Click to select the point at p1 in Figure 1-49 using the Endpoint Object Snap.)

(ADD mode)Specify next point or [Arc/Length/Undo/Total] <Total>: (Press Enter to obtain the total.)

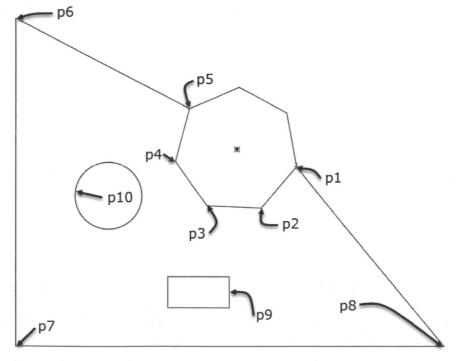

Figure 1-49

Area = 20.1276, Perimeter = 22.5654

Total area = 20.1276
Specify first corner point or [Object/Subtract area/eXit]: **S** (Type **S** to subtract.)

Specify first corner point or [Object/Add area/eXit]: **O** (Type **O** to select objects for subtraction.)

(SUBTRACT mode) Select objects: (Select the rectangle at p9 as shown in Figure 1-50.)

Area = 0.5000, Perimeter = 3.0000 (Area and perimeter of the rectangle.)
Total area = 19.6276 (Total as shown in Command Window when the rectangle is subtracted.)
(SUBTRACT mode) Select objects: (Select the circle at p10 as shown in Figure 1-50.)

Area = 0.8953, Circumference = 3.3543 (Area and circumference of the circle.)
Total area = 18.7322 (Grand Total as shown in Command Window.)
(SUBTRACT mode) Select objects: (Press Enter to end selection.)

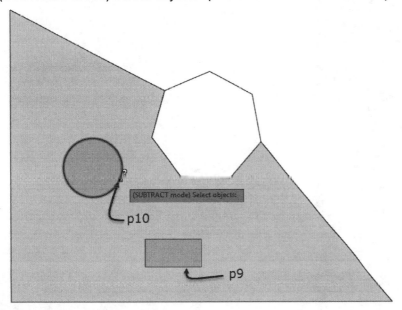

Figure 1-50

1.1 Learning the AutoCAD 2024 Interface Tutorial

The following tutorial provides exercises in the identification of the AutoCAD workspace components.

1. Open AutoCAD 2024 from the Desktop shortcut.
2. From the Start tab, choose the Templates drop-down arrow shown in Figure 1-51 at p1 and verify that the template is set to acad.dwt. If not listed, click **Browse templates...** and select it there first. Left click New as shown in Figure 1-51 at p2.

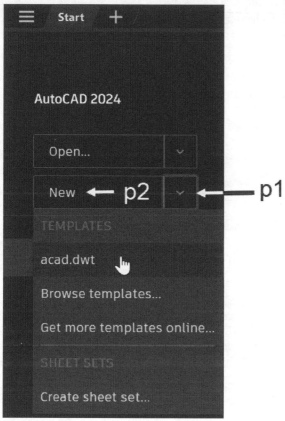

Figure 1-51

3. Choose the Workspace Setting flyout located in the Status bar; verify Drafting & Annotation is checked as shown Figure 1-52.

Figure 1-52

4. To view the name of the workspace in the Quick Access toolbar, as previously shown, click the flyout at p1 shown in Figure 1-53. Select the Workspace option from the flyout as shown at p2.

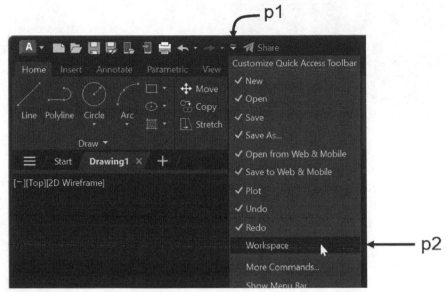

Figure 1-53

5. The toggles of the Status bar can be customized. List the drawing units by selecting the Customization button located in the lower right corner of the graphics window and shown at p1 in Figure 1-54. Left click the Customization button to display the toggle list and choose the Units toggle of the flyout menu shown at p2.

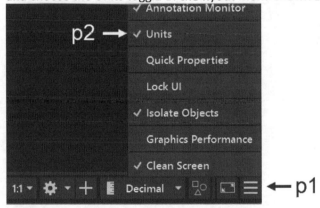

Figure 1-54

6. Verify the Units are set to Decimal as shown in Figure 1-55 in the Status bar. You may click the flyout shown at p1 to change the units.

Figure 1-55

7. The Limits command allows you to set the size of the drawing window. Type **Lim** in the command line and choose the Limits command from the AutoComplete list (Figure 1-56).

Figure 1-56

8. Respond to the workspace prompts as follows:

Command LIMITS

Reset Model space limits:

Specify lower left corner or [ON/OFF]<0.0000,0.0000>: (Press **ENTER** to accept the default values.)

Specify upper right corner <12.0000,9.0000>: (Press **ENTER** to accept the default values.)

9. Notice that when the grid is displayed, a green vertical line originates from the origin of the UCS icon shown in Figure 1-57. Additionally, a red horizontal line originates from the origin.

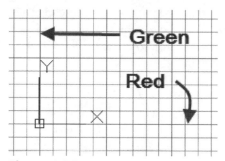

Figure 1-57

10. The current line grid can be changed to a dotted grid; click Snap flyout shown at p1 in Figure 1-58. Click the Snap Settings option to open the Drafting Setting dialog box.

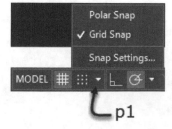

Figure 1-58

11. To display the grid as dots rather than lines check 2D model space in the Grid Style section shown at p1 in Figure 1-59. Choose OK to dismiss the dialog box.

Figure 1-59

12. To return to the line grid, choose **Undo** from the Quick Access toolbar shown at p1 in Figure 1-60.

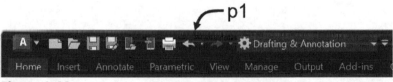

Figure 1-60

13. Choose **SaveAs** command included in the Quick Access toolbar as in Figure 1-61.

Figure 1-61

14. Navigate to your student directory, edit the name to **LAB 1**, and choose **Save** to dismiss the dialog box.

15. You may control the display of the ribbon using the options of its drop-down menu shown at right in Figure 1-62. Click the toggle shown on the left to cycle through various displays of the ribbon.

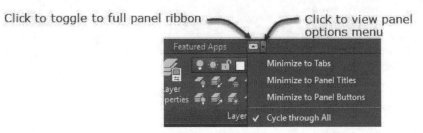

Figure 1-62

16. Choose Minimize to Tabs from the drop-down menu to view the ribbon as shown in Figure 1-63.

Figure 1-63

17. Choose Minimize to Panel Buttons from the drop-down menu to view the ribbon as shown in Figure 1-64.

Figure 1-64

18. Choose the **Cycle through All** option from the drop-down menu at p1 in Figure 1-65. Choose the toggle at left to display the ribbon in each mode and finally to display full panels.

Figure 1-65

19. Choose the **Layout 1** tab located in the lower-left corner of your workspace, as shown in Figure 1-66.

Figure 1-66

20. Right click the Layout 1 tab and choose **Page Setup Manager** from the contextual menu shown in Figure 1-67.

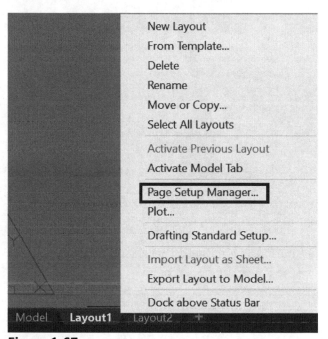

Figure 1-67

21. The Page Setup Manager dialog opens to allow you to specify the printing device and paper size. Click the **Modify** button shown at p1 in Figure 1-68 to open the Page Setup- Layout 1.

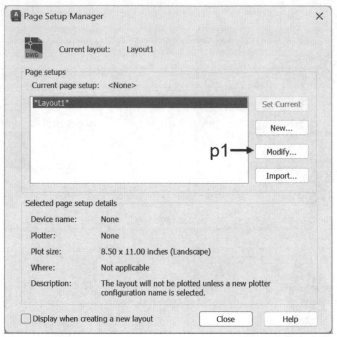

Figure 1-68

22. Click the flyout of the printer name shown at p1 in Figure 1-69 and choose **AutoCAD PDF (General Documentation).pc3**. Click the Paper size flyout shown at p2 and select the **ANSI B (17.00 x 11.00 Inches)**.

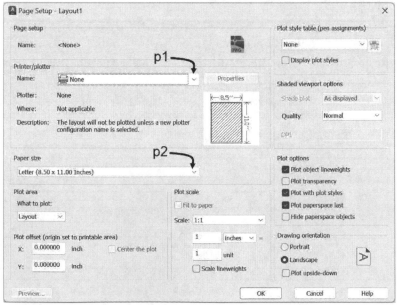

Figure 1-69

23. Verify the Plot area is **Layout 1** and Plot scale is **1:1** as shown in Figure 1-70. Choose **OK** and **Close** to dismiss all dialog boxes.

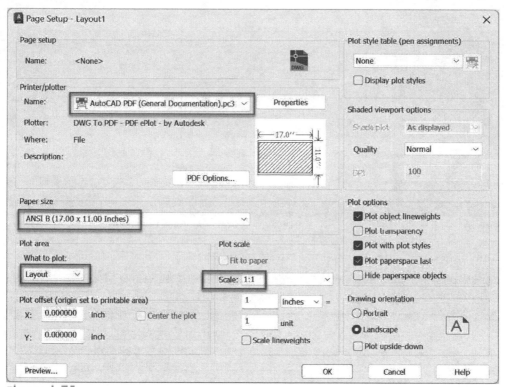

Figure 1-70

24. Note that the Navigation Bar is shown in Figure 1-71 at the far right, and the Paper Space icon is displayed at the far left. Select the viewport boundary as shown below to display the grips for the viewport. Select the grip at p1 and drag the grip to the location near p2. Select the grip at p3 and drag the grip to the location near p4 to resize the viewport to within the dashed line or printer limits for the paper defined in the Page Setup Manager.

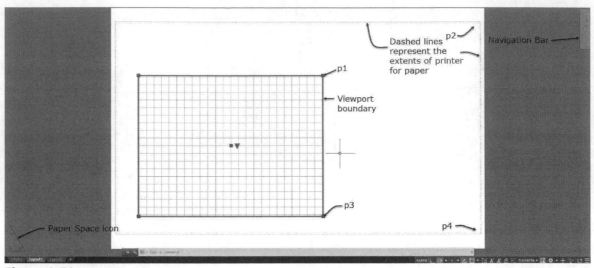

Figure 1-71

25. Double click inside the viewport to activate model space and view the Navigation Bar and View Cube displayed at right within the viewport as shown in Figure 1-72.

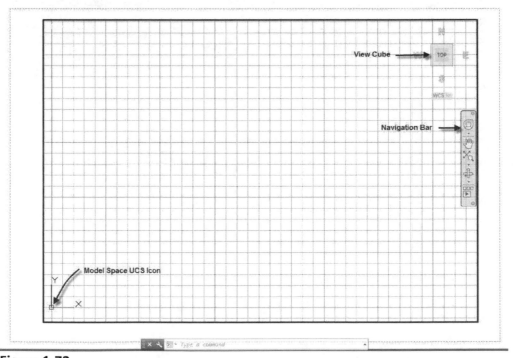

Figure 1-72

26. Choose the Model tab to toggle back to the Model Space, as shown in Figure 1-73.

Figure 1-73

27. Choose Snap Settings of the Snap Mode flyout shown in Figure 1-74.

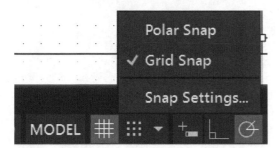

Figure 1-74

28. Edit each of the following settings, press the Tab key to change focus for the setting: Snap X spacing = **.25**, Snap Y spacing = **.25**, Grid X spacing = **.25**, Grid Y spacing = **.25**, Major Grid = **4** and check **Snap On** as shown in Figure 1-75. Click **OK** to dismiss the Drafting Settings dialog box.

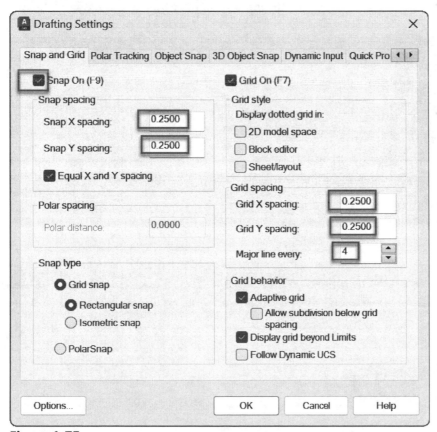

Figure 1-75

29. Choose **Zoom All** from the Zoom flyout of the Navigation Bar, as shown in Figure 1-76.

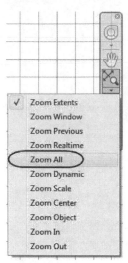

Figure 1-76

30. Dynamic Input allows you to see prompts and enter your response at your cursor. To display the Dynamic Input toggle its display in the Status bar, choose the Customization toggle at the far right of the Status bar and select **Dynamic Input** from the flyout list, as shown in Figure 1-77.

Figure 1-77

31. Choose the Snap Mode in the Status bar as shown at p1 in Figure 1-78 to snap to the grid when drawing a line (it should be toggled on). Note that you can check the snap mode settings by choosing the flyout shown at p2 in Figure 1-78. Verify Dynamic Input is toggled ON in the Status bar as shown at p3 in Figure 1-78.

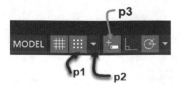

Figure 1-78

32. Right-click Dynamic Input toggle of the Status bar; choose **Dynamic Input Settings**. Verify that each of the checkboxes is checked, as shown in Figure 1-79. Choose **OK** to dismiss the Drafting Settings dialog box.

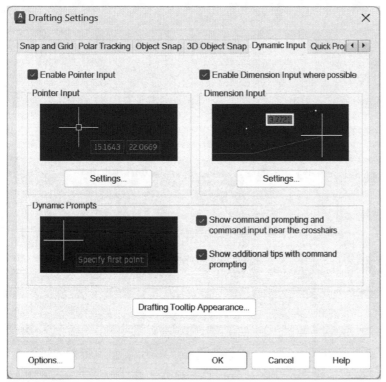

Figure 1-79

33. Choose the Home tab of the ribbon shown in Figure 1-80. Choose the **Line** command from the Draw panel.

Figure 1-80

34. Move the cursor to a grid line and begin drawing a line. Notice that the line snaps to each grid line. Move the cursor up to the second grid to create a ½" vertical line, as shown in Figure 1-81.

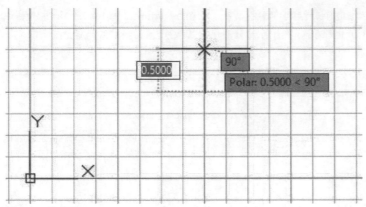

Figure 1-81

35. Draw the following shape shown in Figure 1-82, creating steps with ½" vertical and ½" horizontal components.

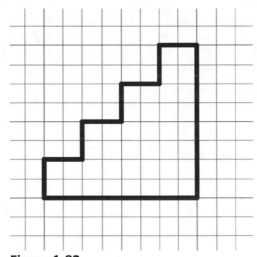

Figure 1-82

36. Click **Zoom Out** from the Zoom flyout of the Navigation bar. Continue to choose the Zoom Out command until you view the UCS icon and the green and red axes, as shown in Figure 1-83.

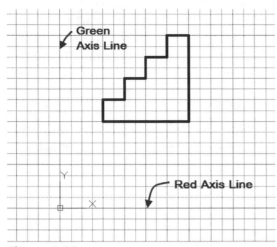

Figure 1-83

37. Choose **Layout 1** tab. In the Status bar, toggle off the Grid and choose 1:2 scale from the Viewport Scale flyout as shown in Figure 1-84.

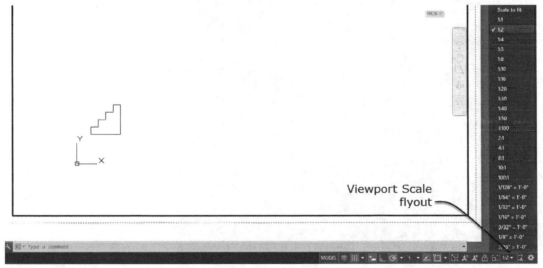

Figure 1-84

38. Choose **Save** from the Quick Access toolbar.

1.2 Using Area and Distance tools Tutorial

In this tutorial you will use the measure tools to determine the area, in square feet, of a building site, subtract the area of the house, and determine the diagonal distance between two corners of the building site.

1. Open the provided **Inquiry.dwg** drawing file located in the AutoCAD 2024 Certified User Exercise Files \ Ch 1 folder (see inside front cover of this book for download instructions).

2. The drawing units are currently set to Decimal. In the next step you will change the units to Architectural to determine distances in feet and areas in square feet. Click at p1 to open the Application Menu, as shown in Figure 1-85. Choose **Drawing Utilities > Units** from the Application menu as shown in Figure 1-85.

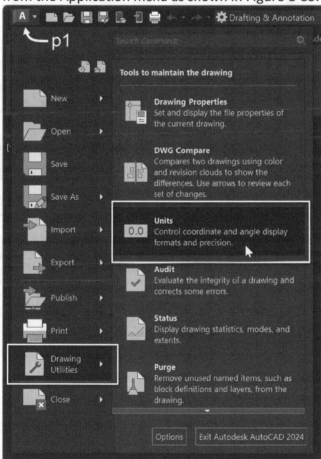

Figure 1-85

3. Choose Architectural from the Length Type drop down of the Drawing Units dialog box as shown in Figure 1-86. Choose OK to dismiss the Drawing Units dialog box.

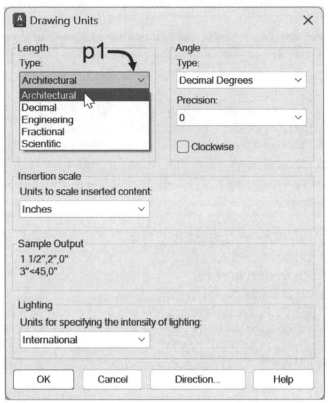

Figure 1-86

4. Choose the **Area** tool from the Measure flyout of the Utilities panel in the Home tab as shown in Figure 1-87.

Figure 1-87

5. Determine the area of the building lot and subtract the area of the house rectangle using the Area command. Refer to Figure 1-88 to specify points, as shown in the following workspace prompts (be sure to follow the steps as written).

Specify first corner point or [Object/Add area/Subtract area/eXit] <Object>: **A** (Type A to specify Add area option.)

Specify first corner point or [Object/Subtract area/eXit]: Click the Endpoint at **p1**.

(ADD mode) Specify next point or [Arc/Length/Undo]: Click the Endpoint at **p2**.

(ADD mode) Specify next point or [Arc/Length/Undo]: Click the Endpoint at **p3**.

(ADD mode) Specify next point or [Arc/Length/Undo]: Click the Endpoint at **p4**.

(ADD mode) Specify next point or [Arc/Length/Undo/Total] <Total>: Click the Endpoint at **p1**.

(ADD mode) Specify next point or [Arc/Length/Undo/Total] <Total>: Press **Enter** to view the Total Area in square feet as shown in bold below.

Area = 2342245.36 square in. (**16265.5928 square ft**.), Perimeter = 518'-11 5/8"

Total area = 2342245.36 square in. (16265.5928 square ft.)

Specify first corner point or [Object/Subtract area/eXit]: **S** (Press S to choose the Subtract option.)

Specify first corner point or [Object/Add area/eXit]: **O** (Press O to choose the Object option.

(SUBTRACT mode) Select objects: (Select the rectangle representing the house at **p5**. The Area and the Total Area is shown in bold below.)

Area = 274176.00 square in. (**1904.0000 square ft**.), Perimeter = 192'-0"

Total area = 2068069.36 square in. (**14361.5928 square ft**.)

(SUBTRACT mode) Select objects: (Press **Escape** to end the command.)

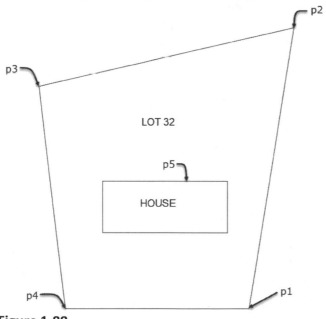

Figure 1-88

6. Choose the **Distance** command from the Measure flyout in the Utilities panel. Respond to the workspace prompts shown below to measure the distance from p1 to p2, as shown in Figure 1-89.

Command: MEASUREGEOM
Enter an option [Distance/Radius/Angle/ARea/Volume/Quick/Mode/eXit] <Distance>:
_distance
Specify first point: Select the endpoint as shown at **p1** in Figure 1-89.
Specify second point or [Multiple points]: Select the endpoint as shown at **p2** in Figure 1-89. (The distance is shown in bold below.)

Distance = **166'-6 11/16**", Angle in XY Plane = 133, Angle from XY Plane = 0
Delta X = -114'-6 1/8", Delta Y = 120'-11 3/8", Delta Z = 0'-0"

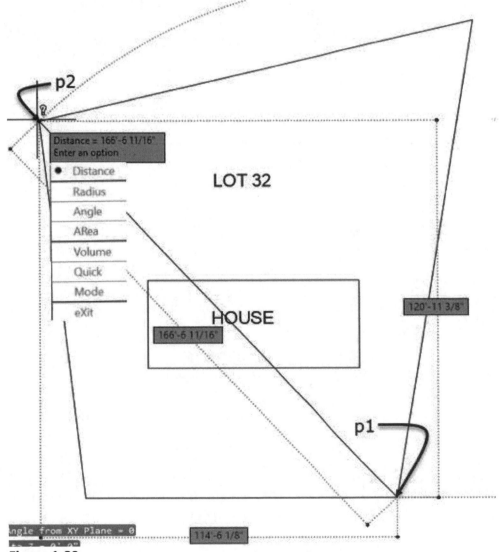

Figure 1-89

1.3 Modifying and Creating a Template file Tutorial

In this tutorial, you will open a customized template file provided by the publisher, make changes, and save the edited file in your student directory. In addition to editing a template file, you will convert an AutoCAD (.dwg file extension) drawing file into a template file (.dwt file extension) using the SaveAs command.

1. Choose **Open** from the Quick Access toolbar to display the Select File dialog box. Choose the **Drawing Template (*dwt)** options from the Files of type flyout as shown at p1 in Figure 1-90.

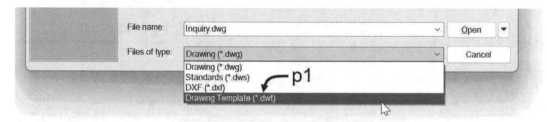

Figure 1-90

2. When you select the Drawing Template filter, the directory switches to the AutoCAD Template folder as shown in Figure 1-91.

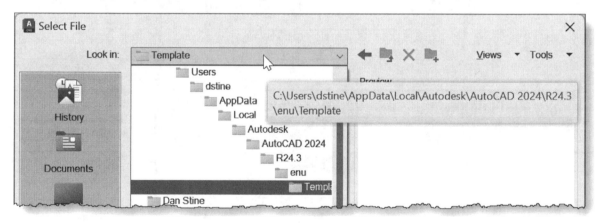

Figure 1-91

3. Navigate to the AutoCAD 2024 Certified User Exercise Files \ Exercise Files\ Ch1 \ Arch Civil Templates and choose the **Detailing Template.dwt** file as shown in Figure 1-92. Choose **Open** to dismiss the dialog box.

Figure 1-92

4. Change the **Units** from Engineering to Architectural. Edit the Length Type and Precision as shown in Figure 1-93.

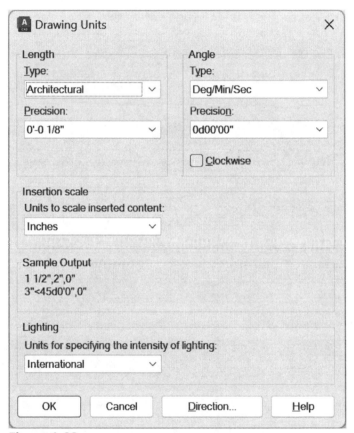

Figure 1-93

5. Choose the Annotate tab; notice that the current text style, dimension style, and leader styles have been edited to include the annotative name, as shown in Figure 1-94.

Figure 1-94

6. Determine the Limits for the drawing and record the coordinates below:

7. X = _____ Y = _____ Lower Left Corner
   X = _____ Y = _____ Upper Right Corner.

8. Choose **SaveAs** from the Quick Access toolbar and save the new template in your student folder as **"your name" Arch Detailing Template**. Close the new template file by clicking the X on the drawing tab.

9. In the remaining steps you will open a civil drawing and create a civil template from the AutoCAD.dwg file.

10. Choose **Open** from the Quick Access toolbar. Edit the **Files by type** filter from *.dwt to *.dwg. Navigate to the AutoCAD 2024 Certified User Exercise Files \Exercise Files \ Ch 1\ Arch Civil Templates folder as shown in Figure 1-95. Select the **Civil Drawing.dwg** file. Choose **Open** to dismiss the Select File dialog box.

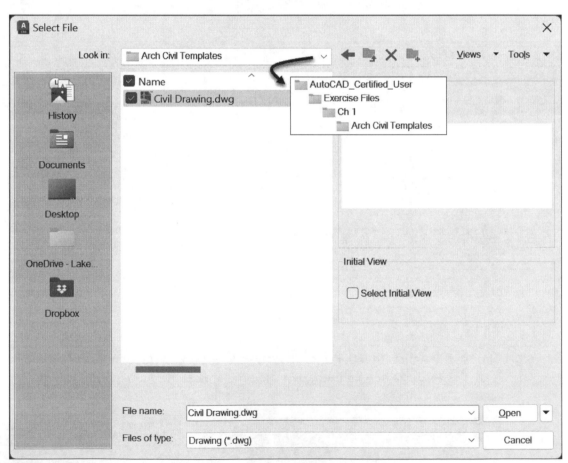

Figure 1-95

11. Determine the Length type and precision for the units of the drawing. Record the Length units below.

_____

12. Choose **SaveAs** from the Quick Access toolbar. Edit the **Files by type** filter to AutoCAD Drawing Template (*.dwt) as shown at p1 in Figure 1-96. Navigate the folder location to your student folder shown at p2. Edit the name to **"your name" Civil Template** as shown at p3 in Figure 1-96. Choose **Save** from the Save Drawing As dialog box. When the Template Options dialog box opens choose **OK** as shown at p4 to accept the settings and Close the file.

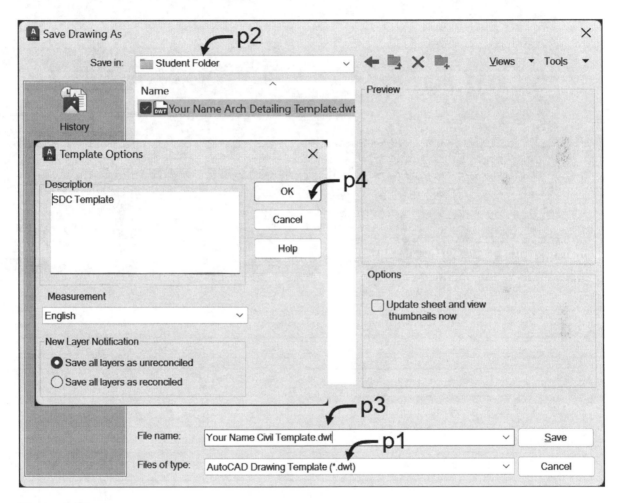

Figure 1-96

1.4 AutoCAD Interface Components Quiz

Write the name of each user interface (UI) component identified below to test your knowledge.

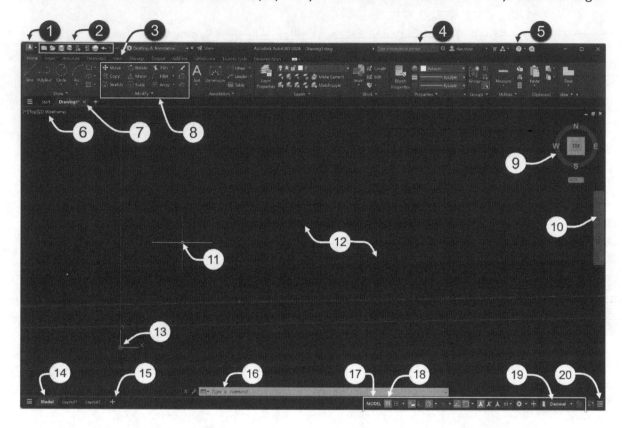

1. _____   11. _____

2. _____   12. _____

3. _____   13. _____

4. _____   14. _____

5. _____   15. _____

6. _____   16. _____

7. _____   17. _____

8. _____   18. _____

9. _____   19. _____

10. _____   20. _____

Chapter 2
Drawing Objects

This chapter will review the basics of drawing, including coordinate entry methods and creating basic entities with an emphasis on Direct Distance Entry, Dynamic Input, and Polar Tracking drawing techniques. Following the review material, there are two tutorials to help you prepare for the test.

The student will be able to:
1. Draw lines using the following coordinate entry methods: absolute coordinates, relative coordinates, relative polar and direct distance entry.
2. Create entities such as lines, arcs, circles, ellipses, rectangles, and polygons using dynamic input and polar tracking tools.
3. Create polygons using the following methods: Inscribed in circle, Circumscribed about circle and Edge.

Coordinate Input Methods

The four methods of inputting locations in AutoCAD are absolute coordinate, relative coordinate, relative polar, and direct distance. The most frequent method used on the test is direct distance. All methods reference the Cartesian coordinate system of horizontal axis of X values and the vertical axis of Y values.

Absolute Coordinates

When you enter values for a line using the absolute coordinates you must toggle off Dynamic Input since dynamic input applies relative polar. In Figure 2.1, the line segment shown started at 0,0 and ended at 2,2.

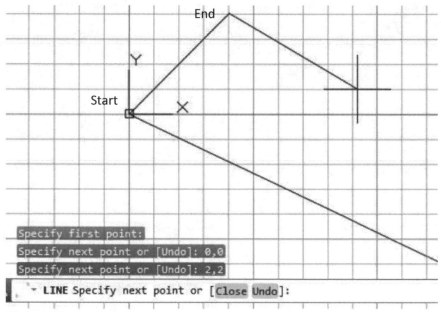

Figure 2.1

Relative Coordinates

When you specify a line using relative coordinates with dynamic input off and type the @ key, the relative mode is turned on. In Figure 2-2, the first point of the line was specified at the absolute coordinate of **5,1** and the next point using relative coordinates typing **@2,0**. If Dynamic Input is toggled on, the prompts at the crosshairs are for direct distance entry; however, when you type **@** and the comma, the prompt changes to the relative coordinates.

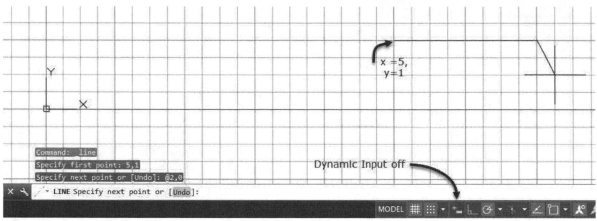

Figure 2.2

Relative Polar Entry

Relative Polar is the default system in place when Dynamic Input is on; therefore, you type the distance and press Tab to type the angle. If Dynamic Input is off, type the @ key to toggle relative and enter the distance followed by the < key then enter the angle.

Direct Distance Entry

Direct Distance entry applies relative polar for the distance and you press Tab to type the angle or just use the mouse to set the angle. The angle can be controlled by Ortho or Polar Tracking. If you toggle polar tracking on, ortho is turned off. The Polar Tracking flyout menu includes preset angles shown In Figure 2-3. Select Track Setting... from the flyout menu to open the Drafing Settings dialog box, set to the Polar Tracking tab, as shown in Figure 2-4.

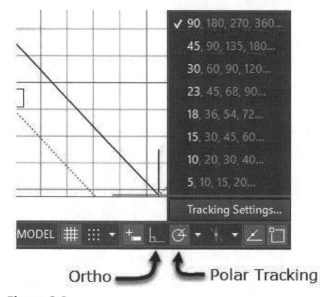

Figure 2.3

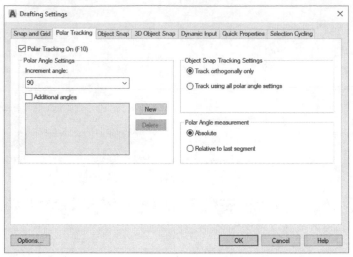

Figure 2.4

Drawing Basic Shapes

The basic shapes of the Draw panel on the Home tab of the ribbon include the Line, Circle, Arc, Rectangle, Polygon, and Ellipse commands. The tutorial at the end of the chapter includes an example of each basic shape using the ribbon. The following is a summary of how to access the command and the workspace prompts.

Line

Choose the Line command from the Home tab, Draw panel as shown in Figure 2-5.

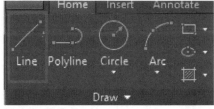

Figure 2-5

Although you may have used this command, try the following description that uses Dynamic Input. Toggle ON Dynamic Input and Polar Tracking in the status bar to show the workspace prompts at the crosshairs. Respond to the workspace prompts as shown below.

Command: _line
Specify first point: (Overtype **1,1** and press **Enter** in the dynamic input fields as shown at p1 in Figure 2-6. Move the crosshairs to a location near p2.)

Specify next point: (Overtype **2** in the dynamic dimension as shown at p3 in Figure 2-6. Press Tab to move focus to p4. Overtype **0** in the dynamic dimension as shown at p4 in Figure 2-6.)

Specify next point: Press Enter to end the command and view the line as shown at p5 in Figure 2-6.

Note: If you press Enter twice or the Space bar, you will repeat the command.

Press Enter three times to end the command, repeat the command, and lock the crosshairs to the endpoint of the last line segment.

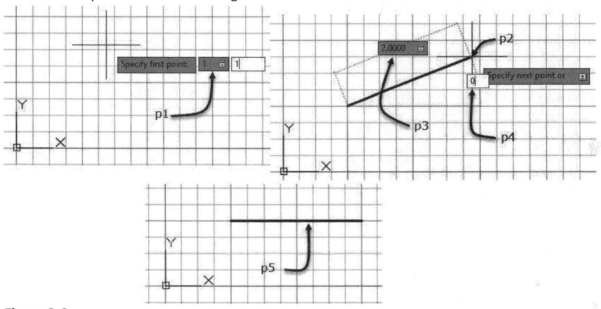

Figure 2-6

Circle

The Circle command is located on the Draw panel in the Home tab as shown in Figure 2-7. If you choose the Circle command, you are prompted to specify the Center and the radius to create a circle. If you select an option from the flyout this command option stays on top for the Circle command. The flyout of the Circle command includes options to create a circle through 2 points and three points without specifying a center. Two additional options are available to create a circle using the tan object snap to create the points that define a circle.

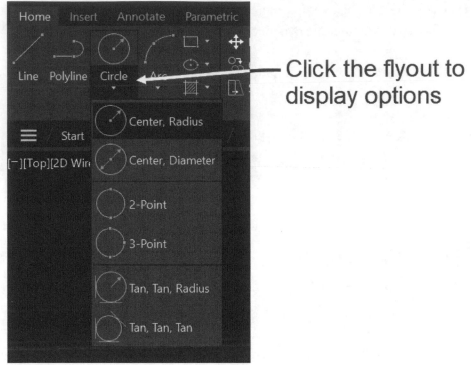

Figure 2-7

Although you may have used this command, try the following description that creates a circle through 2 points using Dynamic Input. Toggle ON Dynamic Input and Polar Tracking in the status bar to show the workspace prompts at the crosshairs. Choose the Circle flyout to display options, select 2-Point, and respond to the workspace prompts as shown below.

> Command: _circle
> Specify first end point of circle's diameter: (Overtype **1,1** and press Enter in the dynamic dimension as shown at p1 in Figure 2-8. Note you are entering first end point using absolute coordinates.)
> Specify second end point of circle's diameter: (Overtype **@1,1** and press **Enter** in the dynamic dimension as shown at p2 in Figure 2-8. Note you are entering second end point using relative coordinates.)
> The circle is shown in Figure 2-9.

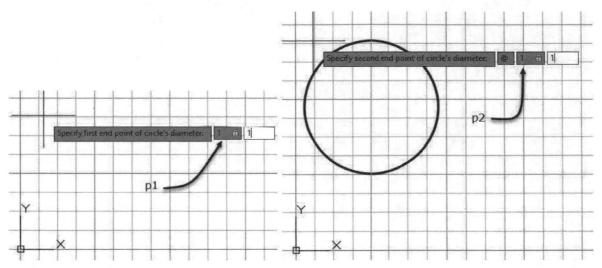

Figure 2-8

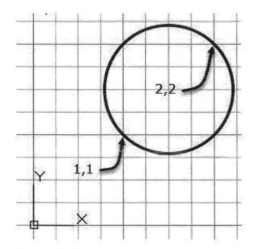

Figure 2-9

Arc

Choose the Arc command from the Home tab, on the Draw panel as shown in Figure 2-10. The Arc button can be preset in any of the eleven methods of placing an arc. Each method presets the prompts. Arcs are drawn counterclockwise from the start point; however, you can press the CTRL key to switch directions.

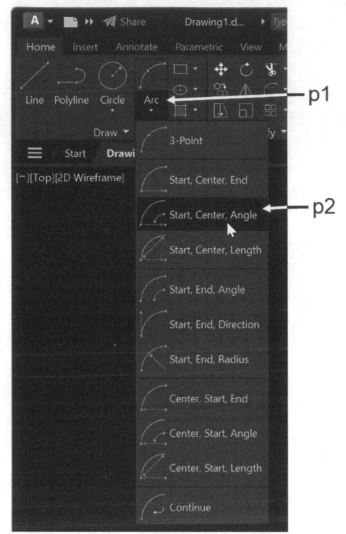

Figure 2-10

Open the **Arc Demo.dwg** drawing file located in the **AutoCAD 2024 Certified User Exercise Files\ Exercise Files \ Certification Demo \ Ch 2** folder to create an Arc as described in the following example using the Start, Center, Angle option of the Arc command as shown at p2 in Figure 2-10.

Command: _arc

Specify start point of arc or [Center]: (Move crosshairs to end of line shown at p1 in Figure 2-11; click when Endpoint object snap marker is displayed to select the endpoint of a line.)

Specify center point of arc: **.5** (Move cursor left, type **.5**, and press **Enter** to locate center as shown at p2 in Figure 2-11.)

Specify included angle (hold Ctrl to switch direction): **90** (Type **90** in polar angle dynamic dimension as shown at p3 in Figure 2-11.)

The ½" radius arc is shown at p4 in Figure 2-11.

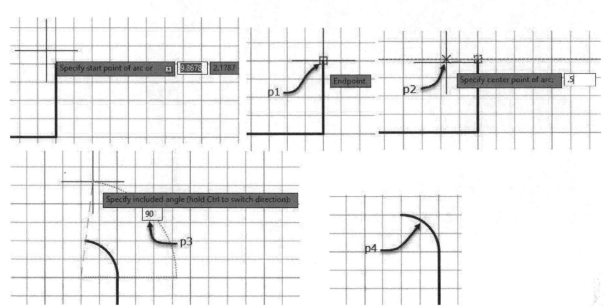

Figure 2-11

Rectangle

Choose the Rectangle command from the Home tab, Draw panel as shown in Figure 2-12. The Rectangle command creates a closed polyline in the shape of a rectangle; therefore, there are options to set Width and apply fillets or chamfers for all corners. The rectangle size can be specified by entering the locations of the diagonal corners of the rectangle or typing the dimensions of width and length.

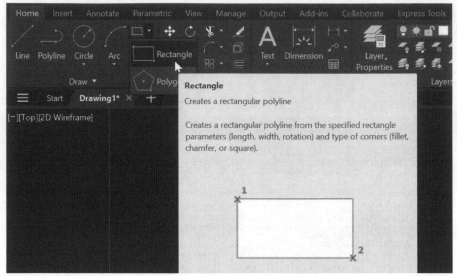

Figure 2-12

Although you may have used this command, try the following description that creates a 2" long and 1" wide rectangle. Open the **Rect Demo.dwg** drawing file located in the **AutoCAD Certified User Exercise Files \ Exercise Files\ Certification Demo \ Ch 2** folder to create a Rectangle as described below. Toggle ON Dynamic Input and Polar Tracking in the status bar. Choose the Rectangle command from the Rectangle flyout, as shown in Figure 2-12, and respond to the workspace prompts as shown below.

Command: _Rectang
Specify first corner point: *(Overtype **.5, .5** and press **Enter** in the dynamic input prompt as shown at p1 in Figure 2-13).*
Move the cursor to a point near p2 as shown in Figure 2-13 to specify the direction to create the rectangle relative to the first corner.
Specify the other corner point or: *(Overtype **2, 1** and press **Enter** in the dynamic input prompt as shown at p3 in Figure 2-13.)*
The rectangle is shown at p4 as shown in Figure 2-13.

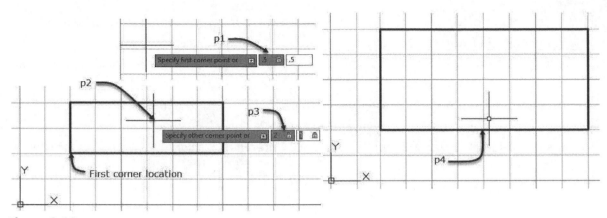

Figure 2-13

Polygon

Choose the Polygon command from the Rectangle flyout in the Home tab, Draw panel as shown in Figure 2-14. There are three options to the command: Inscribed in circle, Circumscribed about circle, and Edge. The Inscribed in circle method creates a polygon inside of a circular shape used to specify the size of the polygon. The Inscribed in circle method specifies the polygon distance across corners. The Circumscribe about circle method sizes the polygon based on the distance across flats since the polygon surrounds the circle. The Edge method sizes the polygon based on the length of a side and the number of sides.

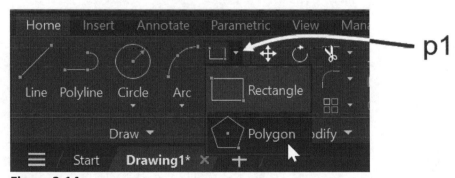

Figure 2-14

Although you may have used this command, try the following description that creates a 5-sided polygon using the Circumscribe about circle using Dynamic Input. Open the **Polygon Demo.dwg** drawing file located in the **AutoCAD Certified User Exercise Files \ Exercise Files\ Certification Demo \ Ch 2** folder to create a Polygon as described below. Toggle ON Dynamic Input and Polar Tracking in the status bar. Choose the Polygon command from the Rectangle flyout and respond to the workspace prompts as shown below.

    Command: _Polygon

Enter number of sides <4>: *(Overtype **5** and press **Enter** in the dynamic input prompt as shown at p1 in Figure 2-15.)*

Specify center of polygon or: *(Overtype **1,1** and press **Enter** in the dynamic input prompt as shown at p2 in Figure 2-15.)*

Enter an option: *(Press the **down arrow key** of the keyboard to select the Circumscribed about circle; press Enter as shown at p3 in Figure 2-15.)*

Specify radius of circle: *(Overtype **.5** and press **Enter** in the dynamic input prompt as shown at p4 in Figure 2-15.)*

Note the polygon is shown in Figure 2-16.

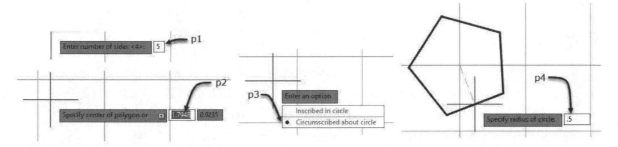

Figure 2-15

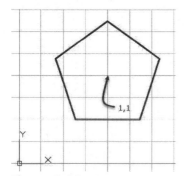

Figure 2-16

Try the following description that creates a 5-sided polygon using the Edge method using Dynamic Input. Toggle ON Dynamic Input and Polar Tracking in the status bar. Choose the Polygon command from the Rectangle flyout and respond to the workspace prompts as shown below.

Command: _Polygon

Enter number of sides <4>: *(Overtype **5** and press **Enter** in the dynamic input prompt as shown at p1 in Figure 2-17.)*

Specify center of polygon or: *(Choose the **down arrow key** twice on the keyboard to display and select the **Edge** option as shown at p2 in Figure 2-17.)*

Specify the first endpoint of edge: *(Overtype **5,.5** and press **Enter** to specify the coordinates as shown at p3 in Figure 2-17.)*

Specify the second endpoint of edge: *(Overtype **1**, press **Tab**, overtype **0** and press **Enter** to specify the length of the edge as shown at p4 in Figure 2-17.)*

The completed polygon is shown at p5 in Figure 2-17.

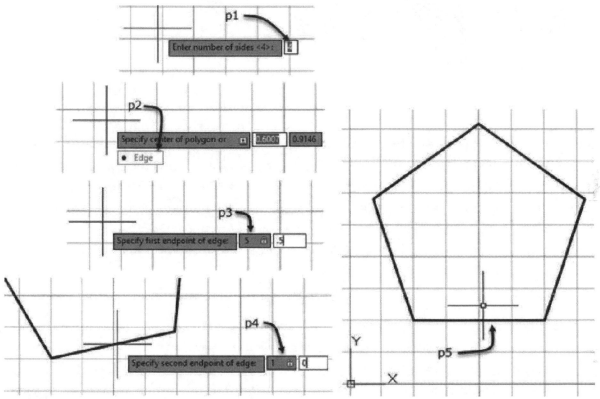

Figure 2-17

Ellipse

Choose the Ellipse command from the Home tab, Draw panel as shown in Figure 2-18. The Ellipse command creates an ellipse by specifying the center and the distances from the center to axis 1 and axis 2 as shown at right below.

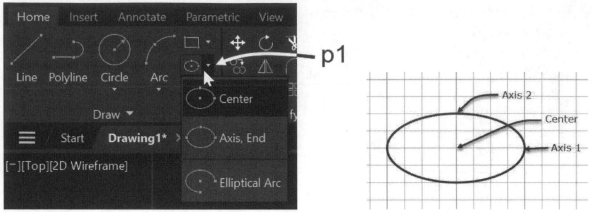

Figure 2-18

Although you may have used this command, try the following description that creates an ellipse using Dynamic Input with a major axis of 1 and a minor axis of .5. Open the **Ellipse Demo.dwg** drawing file located in the AutoCAD Certified User Exercise Files \ Exercise Files\ Certification Demo \ Ch 2 folder to create an Ellipse with a major axis of 1 and a minor axis of .5 as described below. Toggle ON Dynamic Input and Polar Tracking in the status bar to show the workspace prompts at the crosshairs. Choose the Ellipse Center command and respond to the workspace prompts as shown below.

> Command: _Ellipse
> Specify center of ellipse: *(Overtype **3,1** and press **Enter** in the dynamic dimension as shown at p1 in Figure 2-19. Move the crosshairs to the right as shown at p2 in Figure 2-19.)*
> Specify endpoint of axis: *(Overtype **1** in the distance field of the dynamic dimension as shown at p3 in Figure 2-19. Press **Tab** and enter **0** in the polar angle; press **Enter** to specify the major axis)*
> Specify distance to other axis or: *(Type **.5** in the distance field as shown at p4 in Figure 2-19 to specify the minor axis; press **Enter** to complete the command.)*
> The completed Ellipse is shown at p5 in Figure 2-19.

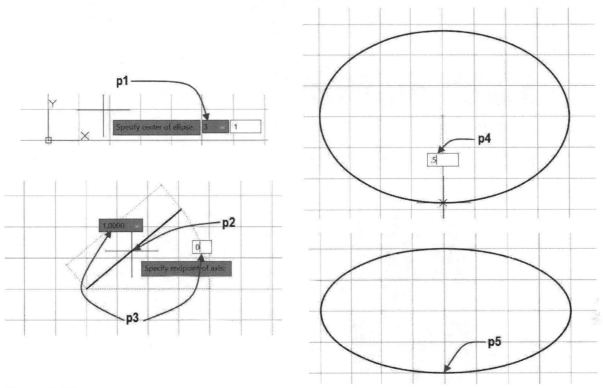

Figure 2-19

2.1 Creating Drawings Using Direct Distance Entry Tutorial

The following tutorial applies the use of Direct Distance Entry to set the angle and direction using the mouse and enter the distance from the start point to draw lines. Press the Tab key to change the focus to a field of dynamic input. Press Shift Tab to change the focus to the previous field.

1. Open AutoCAD from the Desktop shortcut.
2. Choose **New** from the Quick Access toolbar.
3. Choose **SaveAs** from the Quick Access toolbar to save the drawing to your student folder. Name the file **2.1 Direct Distance Your Name.dwg.**
4. Toggle **ON** Polar and Dynamic Input in the Status bar shown in Figure 2-20.

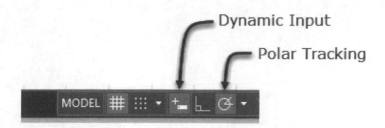

Figure 2-20

5. Right-click the Polar Tracking toggle of the Status bar; choose Tracking Settings as shown in Figure 2-21.

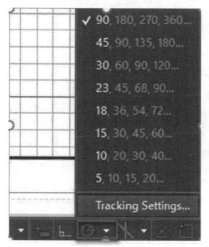

Figure 2-21

6. Edit the Increment angle to **15** degrees as shown below. Verify the remaining settings as shown in Figure 2-22. Choose **OK** to dismiss the dialog box.

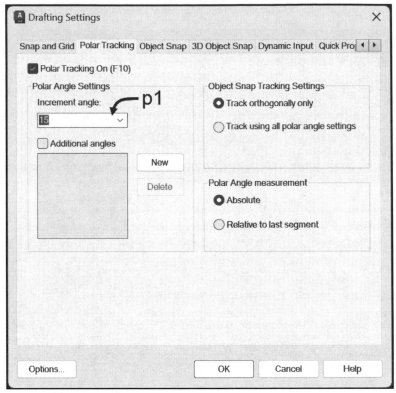

Figure 2-22

7. Choose the Home tab. Choose the **Line** command of the Draw panel as shown in Figure 2-23.

Figure 2-23

8. Respond to the workspace prompts shown in Figure 2-24. Type **2**, press Tab, type **2**, press **Enter** to set the start point to 2,2 using absolute coordinates.

Figure 2-24

9. Move the crosshairs to set the angle to **15** degrees and type **2** in the length field as shown in Figure 2-25, press **Enter**.

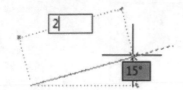

Figure 2-25

10. Move the cursor to set the angle to **90**, type **1** in the length field as shown in Figure 2-26, and press **Enter**.

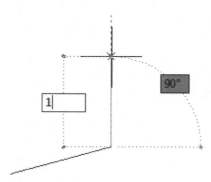

Figure 2-26

11. Move the cursor to the right to display the 0 degree angle, type **.5** in the length field as shown in Figure 2-27, and press **Enter**.

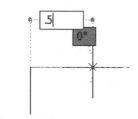

Figure 2-27

12. Move the cursor to display the **45**-degree angle from the horizontal, type **2** in the distance field as shown in Figure 2-28, and press **Enter**.

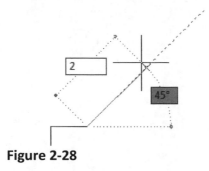

Figure 2-28

13. Continue with the line command to complete the line string as shown in Figure 2-29.

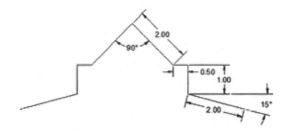

Figure 2-29

14. Verify the Home tab of the Ribbon is current.
15. Right-click **Object Snap** toggle in the Status bar. Choose **Endpoint** to set the Endpoint object snap mode current as shown in Figure 2-30.

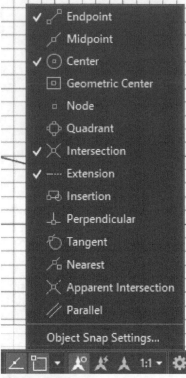

Figure 2-30

16. Choose the **Distance** command from the Measure flyout of the Utilities panel as shown in the Home tab at p1 in Figure 2-31.

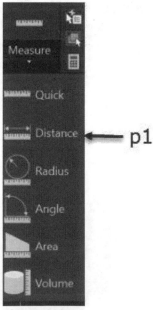

Figure 2-31

17. Select the start and end of the line string created in the previous steps, as shown in Figure 2-32.

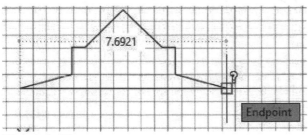

Figure 2-32

18. The distance value is displayed in the workspace. The distance is **7.6921** as shown in Figure 2-33.

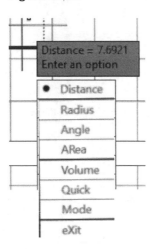

Figure 2-33

19. Press **F2** function key to display the text of the command window.

    Specify first point:

    Specify second point or [Multiple points]:

    Distance = 7.6921, Angle in XY Plane = 0, Angle from XY Plane = 0

    Delta X = 7.6921, Delta Y = 0.0000, Delta Z = 0.0000

20. Now that you have turned on object snap and the endpoint mode, choose the line command and draw 2" lines vertically from the endpoints as shown in Figure 2-34.

Figure 2-34

21. Draw a line connecting the two lines as shown in Figure 2-35.

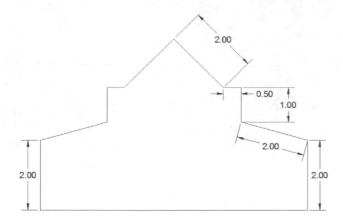

Figure 2-35

22. Choose **Save** from the Quick Access toolbar.

2.2 Drawing Basic Shapes Tutorial

The following tutorial includes practice using Dynamic Input to draw the Circle, Ellipses, Rectangle, Hexagon, Octagon, Square, and other shapes.

1. Choose **New** from the Quick Access toolbar.
2. Choose SaveAs from the Quick Access toolbar to save the file as **2.2 Basic Shapes Your Name** to your student directory.
3. Verify Dynamic Input is toggled **ON** in the Status bar.
4. Choose the **Rectangle** command from the Draw panel of the Home tab as shown in Figure 2-36.

Figure 2-36

5. Draw the rectangle shown in Figure 2-37 using Dynamic Input with the lower left corner located at 1,1 coordinate.

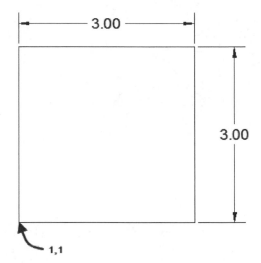

Figure 2-37

6. Draw a rectangle as shown in Figure 2-38 using Dynamic Input with the lower left corner located at 6,1 coordinate.

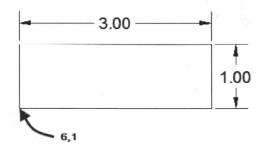

Figure 2-38

7. Choose Center, Diameter tool from the Circle flyout and draw the circle as shown in Figure 2-39 with its center located at 12,2.

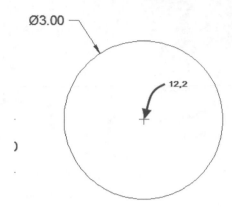

Ø3.00

12,2

Figure 2-39

8. Choose the **Polygon** command from the Rectangle flyout of the Draw panel in the Home tab shown in Figure 2-40.

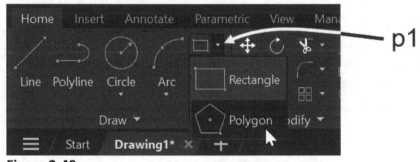

p1

Figure 2-40

9. Draw a hexagon with its center located at **3,7**, as shown in Figure 2-41. Tip: you will use the Circumscribed about circle option of the polygon command.

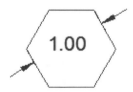

1.00

Figure 2-41

10. Draw an Octagon with its center located at **5,7**, as shown in Figure 2-42. Tip: you will use the Inscribed in circle option.

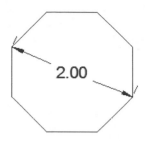

Figure 2-42

11. Draw a pentagon with sides **0.75** in length, with one side located at **7,7**, as shown in Figure 2-43. Tip: you will use the Edge option of the polygon command.

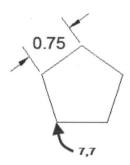

Figure 2-43

12. Choose the Ellipse Center command from the Ellipse flyout of the Draw panel shown at p1 in Figure 2-44.

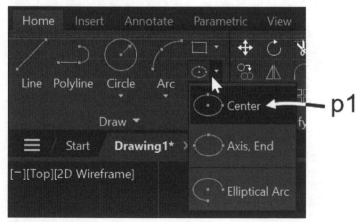

Figure 2-44

13. Draw an ellipse with its center at **10,7**, as shown in Figure 2-45.

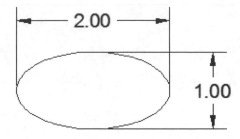

Figure 2-45

14. Draw an arc with its start point at **10,2**, radius equal to **.5**, and an arc angle equal to **90** as shown in Figure 2-46.

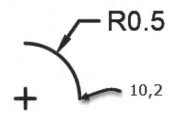

Figure 2-46

15. **Save** and Close the drawing.

Chapter 3
Drawing with Accuracy

This chapter will review the basics of drawing with accuracy, which includes drawing with object snap modes, running and override object snap modes, and object snap tracking techniques. The chapter also includes an overview of techniques to select multiple objects, such as the window, crossing and fence selection using rectangular and irregular polygon shapes. The distance command and the use of Quick Properties to obtain information regarding entities of a drawing are included. Following the review material, there are two tutorials, a quiz, and a drawing assignment to help you prepare for the test.

The student will be able to:
1. Draw lines using running and override object snap modes
2. Identify object snap markers running object snap modes
3. Identify the purpose of object snap modes
4. Customize Quick Properties to determine additional information about entities
5. Identify efficient methods of selecting multiple objects
6. Draw lines and circles located from entities using Object Snap Tracking

Reviewing Object Snaps

Object Snaps are tools available for you to snap your crosshairs to features of an entity such as endpoint or midpoint of a line or arc. The object snap toggle located on the Status shown in Figure 3.1 allows you to turn ON or OFF object snaps and set object snap modes that are active. Object snaps can be Running or active for the drawing session. The modes shown with a check in Figure 3.1 are running.

If you choose the Object Snap Setting shown at p1 you will open the Drafting Settings dialog box and view the running object snap modes as shown in Figure 3.2.

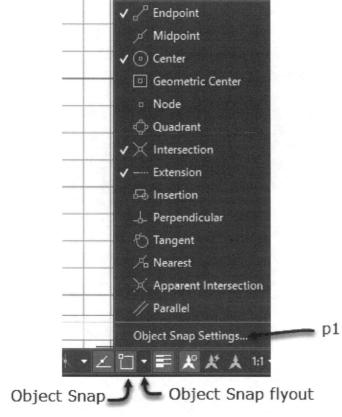

Figure 3-1

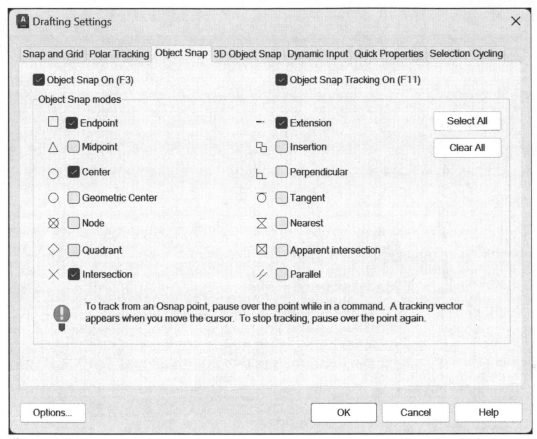

Figure 3-2

An object snap mode can be specifically applied during a command as an override by holding down the shift key or control key and right-clicking to display a shortcut menu, as shown in Figure 3.3. If you select an object snap mode from the menu, only the selected object snap mode will be active.

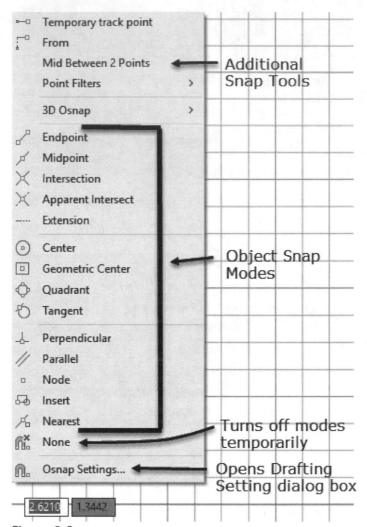

Figure 3-3

The purpose of the object snap modes follows:

> **Endpoint** –snaps to locate the nearest end or corner of a line, arc, or polyline.

> **Midpoint** - snaps to locate the midpoint of the length of a line or arc.

> **Intersection** –snaps to locate the intersection of entities such as lines, arcs, polylines, or circles.

> **Apparent Intersection** –snaps to locate the intersection of two objects that may appear to intersect in the current view if extended.

> **Extension** –creates a temporary extension line from an arc or line, allowing you to locate a point on the extension.

Center –snaps to locate the center of an arc, circle, ellipse, or elliptical arc when you hover over the entity.

Geometric Center –snaps to locate the centroid of a closed polyline or spline such as a polygon or rectangle.

Quadrant –snaps to locate the quadrant points of an arc circle, ellipse, or elliptical arc.

Tangent – locates a tangent point of a line connecting to an arc, circle, ellipse, or elliptical arc.

Perpendicular –snaps to locate a point perpendicular to another entity.

Parallel –constrains a new line segment to be parallel to an existing line that you hover over with your cursor.

Node –snaps to an AutoCAD point, dimension definition point, or dimension text.

Insertion –snaps to the insertion point of an attribute, block, or text.

Intersection –snaps to the intersection of entities such as lines, arcs, polylines, or circles.

Apparent Intersection –snaps to the intersection of two objects that may appear to intersect if extended in the current view.

Nearest –snaps a point nearest the crosshairs on all entities such as lines, arcs, circles, and polylines.

Using Running Object Snap Modes

You use object snaps when you are drawing entities and need to locate an existing entity. If multiple running object snap modes are turned ON, it may be difficult to select the desired marker. If you have chosen the Center and Quadrant running object snap modes described above, when you move crosshairs near the circle, a center marker or one of four quadrant markers may display. You can press the Tab key to quickly toggle to the desired marker location and click to select as shown in Figure 3-4.

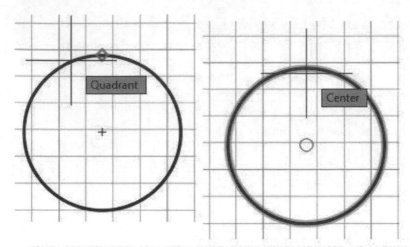

Press Tab key toggle object snap marker locations.

Figure 3-4

Using the Endpoint Object Snap Mode

In Figure 3-5 an existing line was drawn as shown at left. Choose Endpoint Object Snap Mode from the status bar. Choose the Line command; when you move the crosshairs near the right half of the line, a marker for the Endpoint mode is displayed, as shown at p1. When the marker is displayed, left click, and the new line will start at the end of the first line as shown at p2 in Figure 3-5.

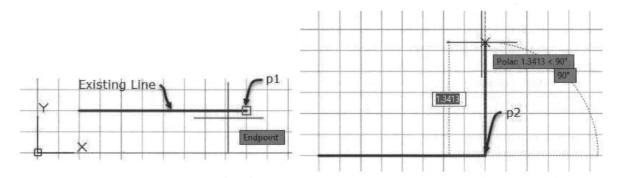

Figure 3-5

Geometric Center Osnap Mode

You can use the Geometric Center Object Snap Mode to draw a circle with the center located at the centroid of the rectangle if the rectangle was created using the Rectangle command (which creates a polyline). Toggle ON Geometric Center as shown at p1 in Figure 3-6. Choose the Circle command Center Radius option, and move the crosshairs along the rectangle to display the Geometric Center marker, shown at p2 in Figure 3-6. Click the green Geometric Center marker to specify the center. Note that if you move the crosshairs, the marker may appear without the green circle; move the crosshairs back to the marker and click when the marker is displayed green.

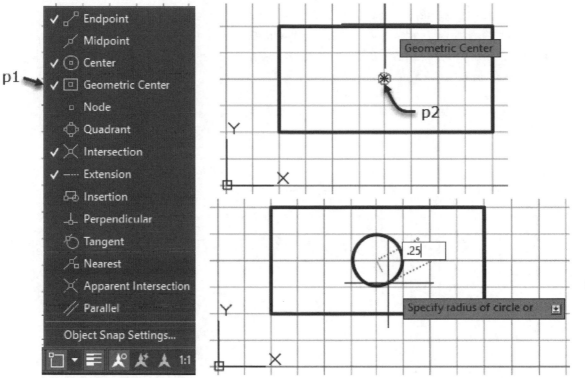

Figure 3-6

Object Snap Tracking

Object Snap Tracking allows you to track from an object snap location on an entity to create an entity with precision. To use object snap tracking, you must turn on Object Snap Tracking and Object Snaps in the Status bar and turn on running object snap mode appropriate to your applications. The following example illustrates the use of object Snap Tracking to create the top view of the part shown in Figure 3-7.

Verify that Endpoint Object Snap Mode is active and Object Snap Tracking is toggled ON in the Status bar.

Choose the Rectang command from the Draw panel in the Home tab.

Command: _rectang

Specify first corner point or: *(Move the cursor over the corner at p1 as shown in Figure 3-7; hover over the Endpoint until the green tracking vector is displayed. Continue to move the cursor up near p2 following the tracking vector, Type 1 in the dynamic dimension field, and press Enter to locate the first corner at p3 as shown in Figure 3-7.)*

Specify the opposite corner: *(Move the cursor over the right corner at p4, and hover over the Endpoint until the green tracking vector is displayed. Continue to move the cursor up following the tracking vector, Type 2 in the dynamic dimension field, and press Enter to locate the opposite corner at p5, as shown in Figure 3-7.)*

The complete rectangle is shown at p6 in Figure 3-7.

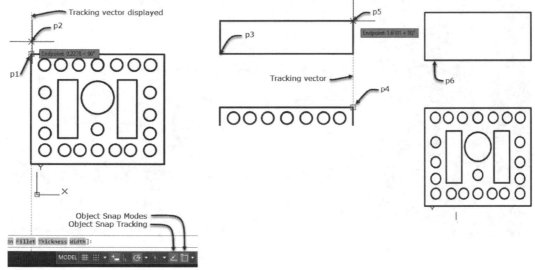

Figure 3-7

Techniques of Selecting Multiple Objects

The following window and crossing selection methods allow you to select objects prior to selecting the editing command; this technique is referred to as noun – verb. When you select objects first, you are not prompted to select objects in the editing commands, which decreases the time required for the edit.

Window rectangular selection- Click to set the cursor for a Window rectangular selection as shown at p1 in Figure 3-8. Move your cursor from p2 to p3; click to select only objects that are entirely enclosed by the rectangular area. Note that the selection must be from **left to right**, and the shape is transparent **blue** in color (by default). To specify the Window rectangular selection in the command line, type **W** at the Select Objects prompt.

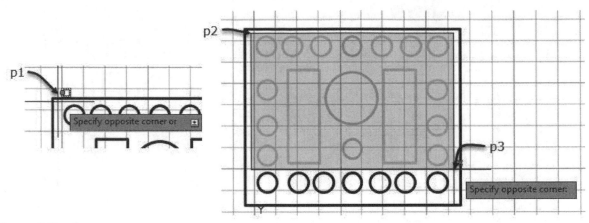

Figure 3-8

Crossing rectangular selection- Click to set the cursor for a Crossing Rectangular selection as shown at p1 in Figure 3-9. Move your cursor from p2 to p3 and click to select objects crossing or enclosed by the rectangular area. Note that the selection must be from **right to left**, and the shape is transparent **green** in color (by default). To specify the Crossing rectangular selection in the command line, type **C** at the Select Objects prompt.

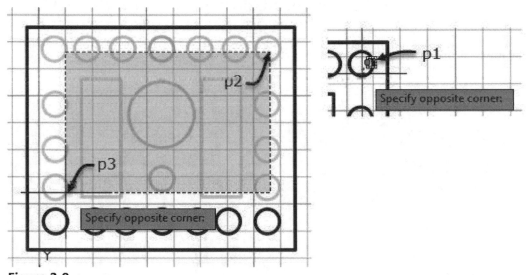

Figure 3-9

Crossing Polygon selection- Click and drag the cursor from p1 shown in Figure 3-10 to the left following the path shown to p2; release the left mouse button to select only objects crossing the irregular shape. The drag technique creates a lasso-crossing polygon. You are prompted in the workspace "Crossing Lasso Press Spacebar to cycle options." This message allows you to press the spacebar toggle to window, fence, or crossing modes for the polygon. Note that the selection must be from **right to left**, and the shape is transparent green in color. To specify the Crossing Polygon selection in the command line, type **CP** at the Select Objects prompt.

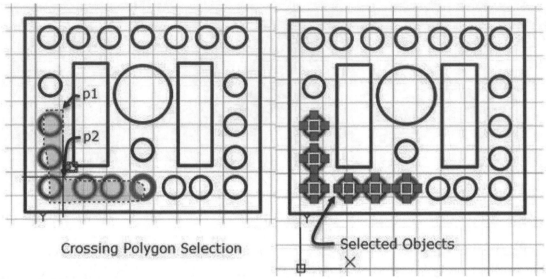

Figure 3-10

Window Polygon selection- Click and drag the cursor as shown in Figure 3-11 from p1 to the right, following the path shown to end at p2. Release the left mouse button to select only objects enclosed in the irregular shape. Note that the selection must be from **left to right**, and the shape is transparent blue in color. To specify the Window Polygon selection in the command line, type **WP** at the Select Objects prompt.

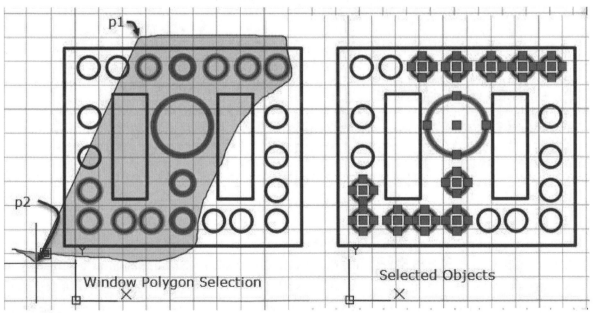

Figure 3-11

Fence- To create a fence line to select all entities that cross the line, type F in the command line and respond to the workspace prompts as shown below:

move (type at the command prompt)

Select objects: **F**

Specify first fence point or pick/drag cursor: (**Select a point** near p1 and drag the cursor to p2, as shown in Figure 3-12.)

Fence Lasso Press **Spacebar** to cycle options: 7 found

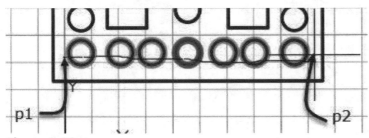

Figure 3-12

Removing and Adding Objects to a Selection Set

To remove objects from a group, hold down the Shift key and select an object for removal. If you are responding to the command line prompts to Select Objects, you can enter an R to select objects to remove or enter an A to add objects.

3.1 Using Object Snap Modes Tutorial

The following tutorial includes methods of selecting entities and applying object snap modes.

1. Click the flyout as shown at p1 to display the **Workspace** in the Quick Access toolbar as shown in Figure 3-13 (if not already displayed).

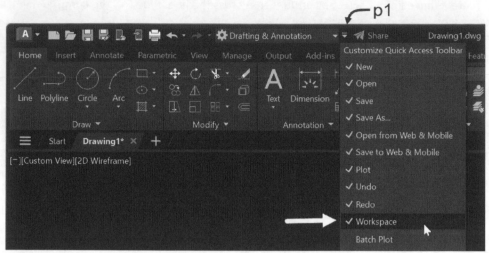

Figure 3-13

2. Verify that **Drafting** & **Annotation** is the current workspace, as shown in the Quick Access toolbar.

3. Choose Open from the Quick Access toolbar.

4. Navigate to the **AutoCAD 2024 Certified User Exercise Files \ Exercise Files \ Ch 3** folder. Choose the **Object Snap.dwg** drawing file**.**

5. Verify **Dynamic Input** is toggled on in the Status bar.

6. Choose the Home tab. Expand the Utilities panel and choose **Point Style** from the Utilities panel shown in Figure 3-14.

Figure 3-14

7. Choose the point style shown at p1 in Figure 3-15. Choose OK to dismiss the Point Style dialog box.

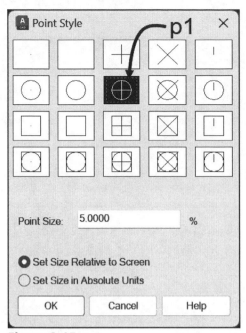

Figure 3-15

8. Toggle ON object snap in the Status bar as shown in Figure 3-16.

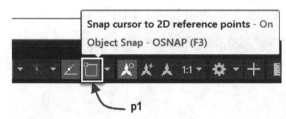

Figure 3-16

9. In this step, you will choose a Running object snap mode from the flyout menu of the Object Snap toggle of the Status bar. Click the flyout shown at p1 in Figure 3-17; choose the **Node** object snap mode.

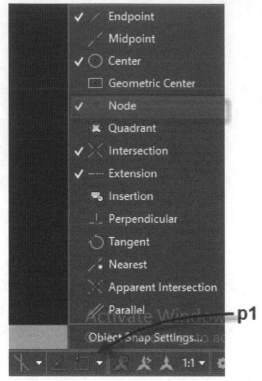

Figure 3-17

10. Verify the Home tab is current on the ribbon. Choose the **Polyline** command shown at p1 in Figure 3-18 from the Draw panel.

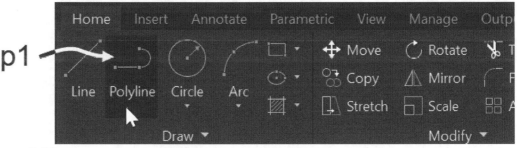

Figure 3-18

11. Draw the polyline through the points shown in Figure 3-19.

Command: _pline

Specify start point: *(Choose the point at p1 as shown in Figure 3-19.)*

Current line-width is 0.0000

Specify next point or [Arc/Close/Halfwidth/Length/Undo/Width]: *(Choose the point at p2 as shown in Figure 3-19.)*

Specify next point or [Arc/Close/Halfwidth/Length/Undo/Width]: *(Choose the point at p3 as shown in Figure 3-19.)*

Specify next point or [Arc/Close/Halfwidth/Length/Undo/Width]: *(Choose the point at p4 as shown in Figure 3-19.)*
Specify next point or [Arc/Close/Halfwidth/Length/Undo/Width]: *(Press Escape to end the command.)*

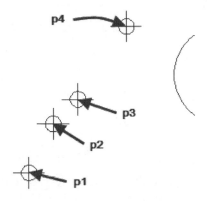

Figure 3-19

12. Click the Customization button of the Status bar to select toggles for display. Click the Quick Properties toggle as shown in Figure 3-20.

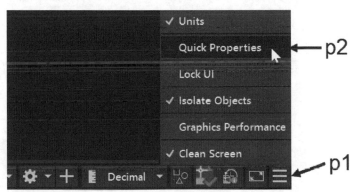

Figure 3-20

13. Verify Quick Properties is toggled ON in the Status bar as shown in Figure 3-21.

Figure 3-21

14. Select the polyline drawn in step 11, to display its grips and open the Quick Properties palette as shown in Figure 3-22.

15. In the following steps, you will customize the content of the Quick Properties palette to add length information for a polyline. Click the **CUI** button at p1 as shown in Figure 3-22, to open the Customize User Interface dialog box shown in Figure 3-23.

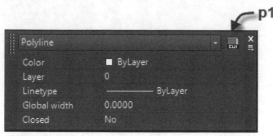

Figure 3-22

16. Verify the Quick Properties is highlighted as shown in Figure 3-23 at p1 and Polyline is selected at p2 in the Customize User Interface dialog box. Check the **Length** box in the Geometry category, as shown at p3 in Figure 3-23.

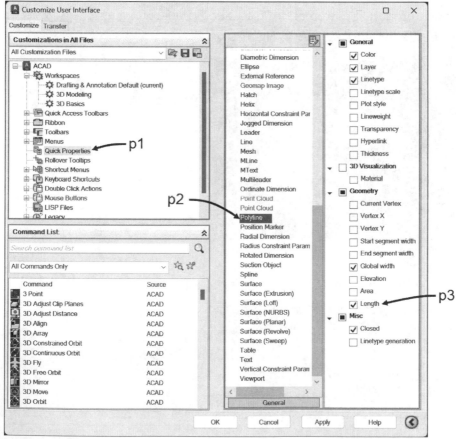

Figure 3-23

17. Choose **OK** to dismiss the Customize User Interface dialog box.

18. Select the Polyline drawn in previous steps to display the Quick Properties palette as shown in Figure 3-24. Note the length of the polyline is listed as 7.2558.

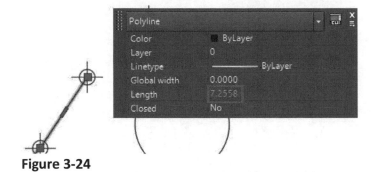

Figure 3-24

19. Press Escape to clear selection and dismiss the Quick Properties palette.
20. In the next series of steps, you will set the object snap modes in the Object Snap tab of the Drafting Settings dialog box.
21. Click the flyout of the Object Snap toggle in the Status bar shown at p1 in Figure 3-25 and choose **Object Snap Settings** to open the Drafting Settings dialog box.

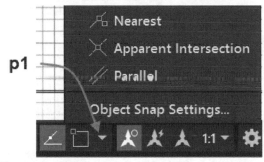

Figure 3-25

22. Choose the **Endpoint**, **Midpoint**, and **Center** object snap modes as shown in Figure 3-26. Verify that **Node** and **Extension** are checked. Choose **OK** to dismiss the Drafting Settings dialog box.

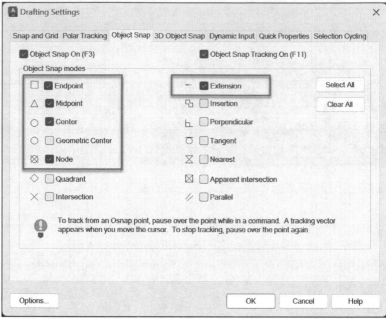

Figure 3-26

23. Choose the **Polyline** command from the Draw panel of the Home tab. Respond to the workspace prompts as follows:

Command: _pline
Specify start point: (Choose the point at **p1** as shown in Figure 3-27 using the Node object snap mode.)
Current line-width is 0.0000
Specify next point or [Arc/Halfwidth/Length/Undo/Width]: (Choose a location at **p2** as shown in Figure 3-27 using the Midpoint object snap mode.)
Specify next point or [Arc/Close/Halfwidth/Length/Undo/Width]: (Choose a point at **p3** as shown in Figure 3-27 using the Midpoint object snap mode.)
Specify next point or [Arc/Close/Halfwidth/Length/Undo/Width]: (Choose a point at **p4** as shown in Figure 3-27 using the Endpoint object snap mode.)
Specify next point or [Arc/Close/Halfwidth/Length/Undo/Width]: (Choose a point at **p5** as shown in Figure 3-27 using the Endpoint object snap mode.)
Specify next point or [Arc/Close/Halfwidth/Length/Undo/Width]: (Choose a point at **p6** as shown in Figure 3-27 using the Midpoint object snap mode.)
Specify next point or [Arc/Close/Halfwidth/Length/Undo/Width]: (Choose a point at **p7** as shown in Figure 3-27 using the Endpoint object snap mode. **Do not end the command; you will continue drawing the polyline in Step 24**.)

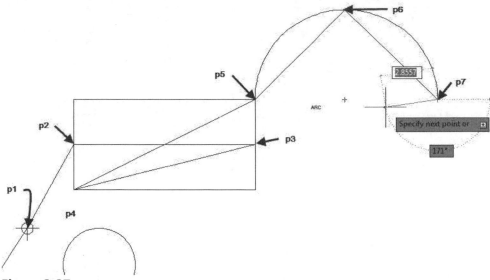

Figure 3-27

24. In this step you will continue the polyline and select the insertion point of the text using the Insert object snap mode from the Override Object Snap shortcut menu. To choose an Override object snap mode, hold down the Shift key, right-click and choose Insert from the Override Object Snap menu shown in Figure 3-28.

    Specify next point or [Arc/Halfwidth/Length/Undo/Width]: _ins of (Move the cursor to a point near **p8** to display the Insert object snap marker of the text; left click to select the Insert point of the text as shown in Figure 3-28.)

    Specify next point or [Arc/Halfwidth/Length/Undo/Width]: (Move the cursor to **p9** to choose the Center of the circle as shown in Figure 3-28.)

    Specify next point or [Arc/Halfwidth/Length/Undo/Width]: (Press **ENTER** to end the command.)

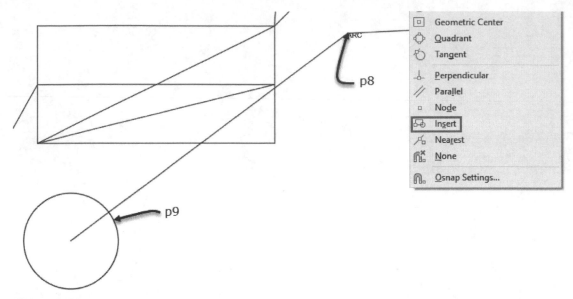

Figure 3-28

25. Select the polyline created in the previous step to display its length in the Quick Properties palette as shown in Figure 3-29. The polyline segment Length should be displayed as 72.4324.

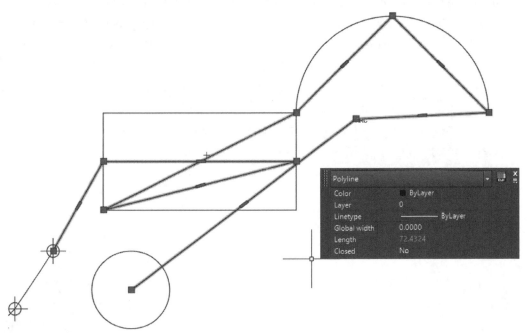

Figure 3-29

26. Save the file as 3.1 Object Snap Modes Your Name in your student folder and Close the drawing.

3.2 Applying Selection Methods Tutorial

The following tutorial includes exercises in the use of the following selection methods: Crossing Polygon, Window Polygon, Fence, Crossing, Window.

1. Verify **Drafting** & **Annotation** is the current workspace as shown in the Quick Access toolbar.
2. Choose Open from the Quick Access toolbar.
3. Navigate to the **AutoCAD 2024 Certified User Exercise Files \ Exercise Files \ Ch 3** folder. Choose the **Selection.dwg** drawing file.
4. Verify Dynamic Input is toggled on in the Status bar. Toggle off Object Snap and toggle ON Polar Tracking in the Status bar.
5. Choose **Copy** from the Modify panel of the Home tab as shown in Figure 3-30.

Figure 3-30

6. Respond to the workspace prompts as shown below.
   Command: _Copy
   Select objects: **wp** *(Type **WP** press **Enter** to specify the Window Polygon method to create a selection set.)*

   First polygon point: <Osnap off> *(Select a point at p1 as shown in Figure 3-31.)*
   Specify endpoint of line or [Undo]: *(Select a point at p2 as shown in Figure 3-31.)*
   Specify endpoint of line or [Undo]: *(Select a point at p3 as shown in Figure 3-31.)*
   Specify endpoint of line or [Undo]: *(Select a point at p4 as shown in Figure 3-31.)*
   Specify endpoint of line or [Undo]: *(Select a point at p5 as shown in Figure 3-31.)*
   Specify endpoint of line or [Undo]: *(Select a point at p6 as shown in Figure 3-31.)*
   Specify endpoint of line or [Undo]: *(Select a point at p7 as shown in Figure 3-31.)*
   Specify endpoint of line or [Undo]: *(Select a point at p8 as shown in Figure 3-31.)*
   Specify endpoint of line or [Undo]: *(Press **Enter** to end the window polygon selection.)*
   5 found
   Select objects: *(Press **Enter** to end object selection.)*

Specify base point or [Displacement] <Displacement>: *(Select a point near p1 as shown in Figure 3-31. Move the cursor to the right.)*

Specify second point or <use first point as displacement>: **15** *(Type **15**, press **ENTER** to copy the selection set 15 inches to the right. Press **Esc** key to end the command.)*

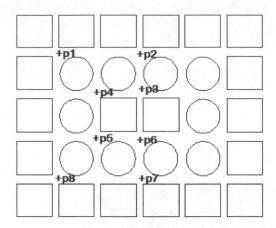

Figure 3-31

7. The selected entities are copied to the right, as shown in Figure 3-32.

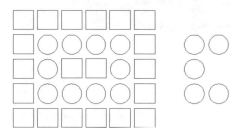

Figure 3-32

8. In the next series of steps, you will select entities using the Fence selection method. Choose **Copy** from the Modify panel of the Home tab.

9. Respond to the workspace prompts as follows:

Command: _copy

Select objects: **F** *(Type **F** to specify the Fence method of selecting objects.)*

Specify first fence point or pick/drag cursor: *(Select a point near p1 as shown in Figure 3-33.)*

Specify next fence point or [Undo]: *(Select a point near p2 as shown in Figure 3-33.)*

Specify next fence point or [Undo]: **ENTER** *(Press **Enter** to end the Fence selection.)*

4 found

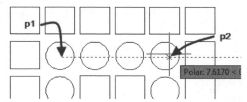

Figure 3-33

Select objects: **ENTER** *(Press **Enter** to end selection.)*

Current settings: Copy mode = Multiple

Specify base point or [Displacement/mOde] <Displacement>: cen of *(Press Shift key and hold down, right click and choose the **Center** object snap mode from the Override Object Snaps shortcut menu.)*

Of *(Move the cursor to the edge of the circumference of the circle shown in Figure 3-34 at p1 below to display the Center object snap marker, left click to choose the **Center** of the circle and define the base point.)*

Specify second point or [Array/Exit/Undo] <Exit>: MID *(Press Shift key and hold down, right click and choose the **Midpoint** object snap mode from the Override Object Snaps shortcut menu.)*

Of *(Move the cursor to the rectangle shown in Figure 3-34 at p2 below to display the Midpoint object snap marker and left click to choose the second point.)*

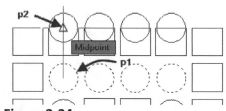

Figure 3-34

10. In the next series of steps, you will copy entities using the crossing window selection method prior to selecting the Copy command.

11. Move the cursor near p1 and click and drag to start the lasso selection; drag the cursor left to create a crossing lasso selection as shown in Figure 3-35. Release the left mouse button at p2.

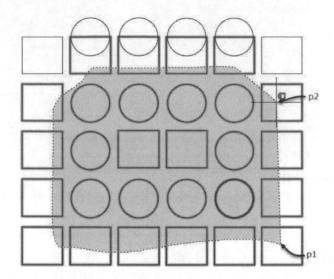

Figure 3-35

12. Choose the **Copy** command from the Modify panel in the Home tab. (Note since you selected objects in the previous steps you are not prompted to select objects when you select the Copy command objects.)

    Specify base point or [Displacement/mOde] <Displacement>: (Select a point near p1 as shown in Figure 3-36.)

    Specify second point or [Array] <use first point as displacement>: (Move the cursor to a point near p2 as shown in Figure 3-36; type **14** in the workspace.)

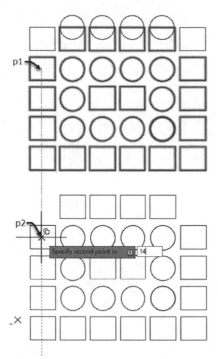

Figure 3-36

13. The objects copied in the previous step are shown in Figure 3-37.

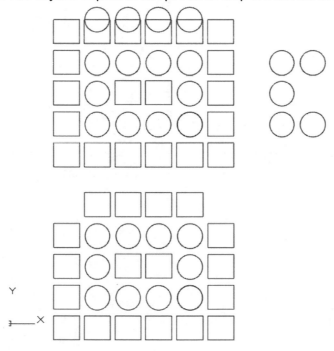

Figure 3-37

14. Choose **SaveAs** from the Quick Access toolbar. Save the file as 3.2 Applying Selection Methods Your Name in your student folder and Close the drawing.

3.3 Object Snap Quiz

Marker	Enter Name of Object Snap Mode
1. ✳	
2. ⊕	
3. ⧗	
4. ⑃	
5. ⊠	
6. ◇	
7. ⊗	
8. ✕	
9. □	
10. △	
11. ⫽	
12. ○	
13. ⅂	
14. ┄	

15. Several running object snap modes have been turned on that apply to the selected entities. What key do you press to display the object snap marker locations without moving the cursor?
    a. Ctrl
    b. Tab
    c. Shift
    d. Enter

3.4 Object Snap Modes and Object Snap Tracking Drawing

Draw the following polygon shown in Figure 2-38 using direct distance entry and object snap tracking. **Do not** use the following commands: **Offset, Extend, Fillet, or Trim**. Draw the shape using only the Circle and Line commands and direct distance entry and object snap tracking.

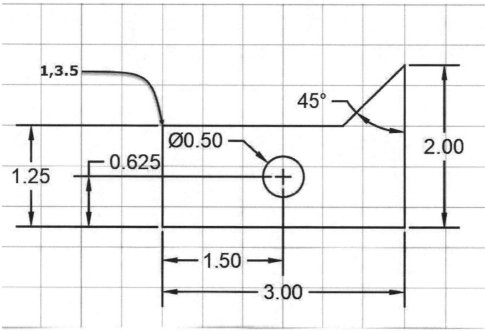

Figure 3-38

Notes:

Chapter 4
Organizing Objects

This chapter will review the basics of using Properties, Quick Select and Layers to organize objects of a drawing. Following the review material, there are two tutorials to help you prepare for the test.

The student will be able to:
1. Use the Properties Palette to determine layer, linetype, and color properties of entities
2. Access and edit a selection set using Quick Select by object type and property
3. Use the Properties palette and ribbon to change the layer properties
4. Create layers with linetype, color, and lineweight
5. Set layer properties per viewport

Using Properties and Quick Select to Organize the Drawing

Designers organize drawings to clearly and efficiently communicate the design intent. AutoCAD includes the Properties palette that allows you to determine the physical properties of one or more objects and to change the layer assignments of selected objects. Often annotation is placed on layers unique from design entities.

Using Properties

To display the Properties palette from the Home tab, choose the launching arrow shown at p1 in Figure 4-1 of the Properties palette or from the View tab choose the Properties button shown at p2 of the Palettes panel in Figure 4-1.

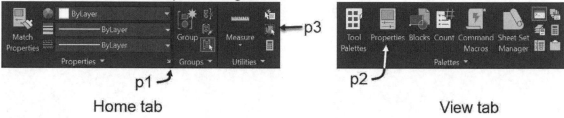

Home tab   View tab

Figure 4-1

Note that you can open the **Quick Select demo** file from the **AutoCAD 2024 Certified User Exercise Files \ Exercise Files \ Certification Demo\ Ch 4** folder to try the technique to create a selection set as you read the following description.

On the test, you could be asked how many circles are in a drawing. To select all the entities of a drawing, choose the **Select All** command located on the Utilities panel of the Home tab shown at p3 in Figure 4-1 to see the total number of entities in the drawing at the top of the Properties palette as shown at p1 in Figure 4-2.

You can also choose the Select button shown at the top of the Properties palette in Figure 4-1 to create a selection set using Window selection and the PSelect command. Note all 40 entities are assigned to layer 0 as shown at p2. Click the flyout to view the quantities per entity type as shown at p3 in Figure 4-2.

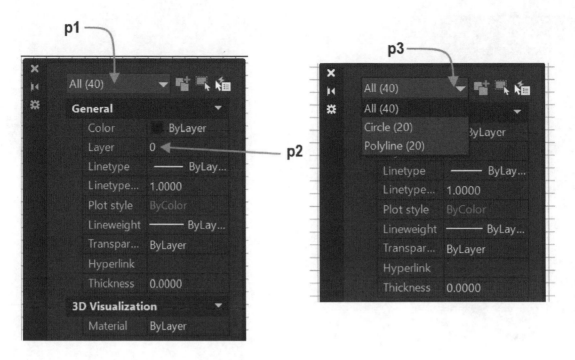

Figure 4-2

If **Circle (20)** is selected from the list (at p3 in Figure 4-2), the listed properties are filtered down to just those elements, even though all 40 elements are still technically selected. The specific information such as Layer, Linetype, and Lineweight is listed as shown at p1 in the General section as shown at p1 in Figure 4-3. If one or more circles have different dimensions, the property will be listed as *VARIES* as shown at p2 in Figure 4-3. Since all circles are located at Z= 0 this property is shown at p3 in Figure 4-3 in the Geometry section.

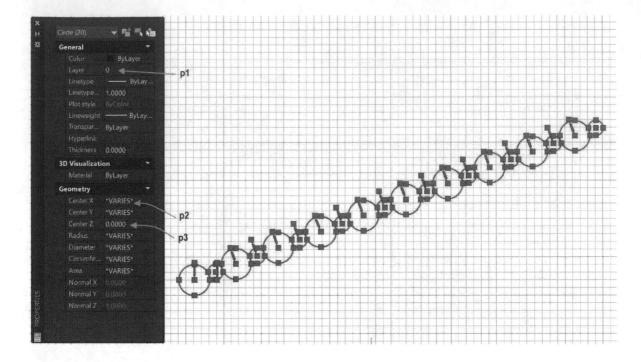

Figure 4-3

Accessing Quick Select

The Quick Select palette allows you to modify a selection set by object type and the properties of all objects of a type. Continue in the **Quick Select demo** file and access the Quick Select palette by choosing the Quick Select button shown at p1 of the Properties palette in Figure 4-4. The Quick Select palette can be chosen from the ribbon as shown at right in Figure 4-4 of the Utilities panel in the Home tab.

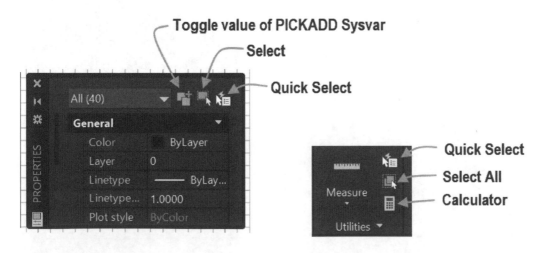

Figure 4-4

The Quick Select palette shown in Figure 4-5 includes multiple object types, such as circles and polylines. Click the flyout shown at p1 to select only one type of object from the list shown at p2 in Figure 4-5. When you edit the object type, the list of properties changes from those common to multiple objects shown at p3 to the properties unique to the selected object type, as shown at p4 in Figure 4-5. You can specify the radius property to equal **.5** to change the selection set to include only circles with the .5 radius. Upon completion of the setting of properties, you can choose the radio button for *Include in new selection set* or *Exclude from the new selection set*, then choose **OK** to dismiss the Quick Select dialog box.

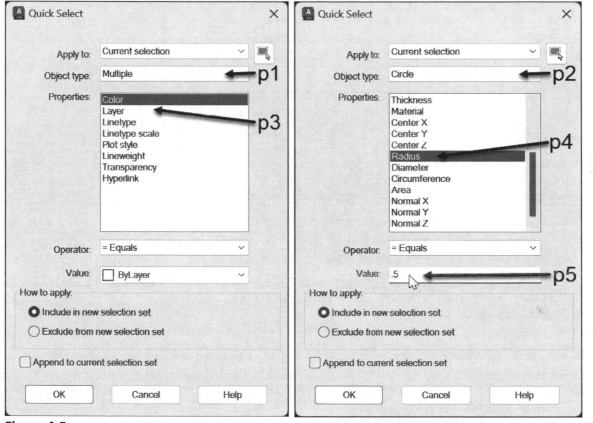

Figure 4-5

Note that the options in the Operator flyout provide an extensive list of available options for searching in a drawing as shown in Figure 4-6.

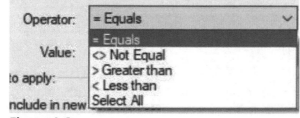

Figure 4-6

Organizing the Drawing Using Layers

Layers are used to separate the content of a drawing. Annotation such as text and dimensions are often placed on unique layers. Design content with similar a function may also be separated by layer. Design content requiring unique color, lineweight, and linetype is placed on layers with the color, lineweight, and linetype assigned by the properties of the layer.

New layers are created, and the properties are specified in the Layer Properties Manager. Typically, the test might ask you to create layers with specific properties, change the layer assigned to an entity, or turn off the display of a layer.

The following are settings for layer control specified in the Layer Properties Manager:
- **On/Off**- toggle on or off display of entities on a layer
- **Freeze/Thaw**- frozen layer content is not displayed or included in regenerations, thaw turns on display of entities on a layer
- **Lock**- entities on locked layers cannot be edited
- **Color**- with bylayer setting all entities of the layer are displayed with the specified color
- **Linetype**- with bylayer setting all entities of the layer are displayed with the specified linetype
- **Lineweight**- with bylayer setting all entities of the layer are displayed with the specified lineweight
- **Plot/NoPlot**- entities on no plot layers will not print or plot
- **Transparency**- with bylayer setting all entities of the layer are displayed with the specified transparency percentage allowing you to see through layer content

Choose the Layer Properties button from the Layers panel of the Home tab of the ribbon, as shown at p1 in Figure 4-7, to open the Layer Properties Manager.

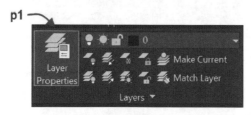

Layers panel of the Home tab

Figure 4-7

The Layer Properties Manager opens as shown in Figure 4-8. The Layer Properties Manager can be resized as other palettes. Each column can be resized by moving the cursor to the edge of the column header to display the two-way arrow and clicking and dragging. Click the New Layer tool from the toolbar as shown in Figure 4-8. A new layer is created; overtype a layer name.

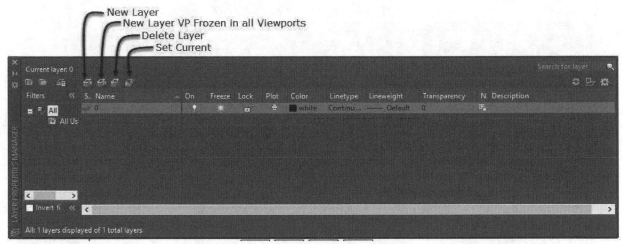

Figure 4-8

If you create a new name you can right click the name and choose Rename Layer from the contextual menu as shown in Figure 4-9.

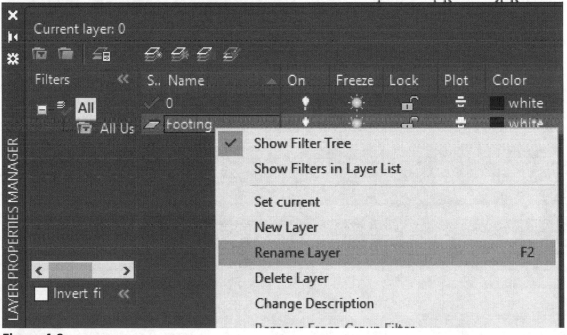

Figure 4-9

Setting Layer Color

To specify color for a layer, click at p1 as shown in Figure 4-10 to open the Select Color dialog box. Select a color as shown at p2 as shown in Figure 4-10. Click OK to dismiss the Select Color dialog box.

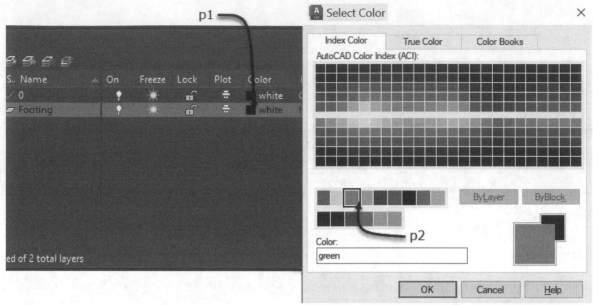

Figure 4-10

Setting Linetype for a Layer

To specify a linetype for a layer, click at p1, shown in Figure 4-11, to open the Select Linetype dialog box. Choose the Load button shown at p2 in Figure 4-11 to load linetypes for your use in the drawing.

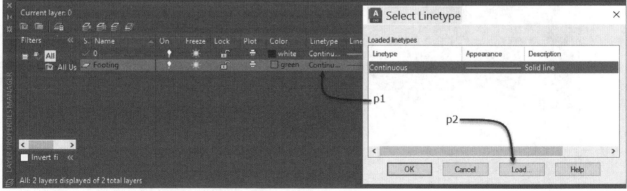

Figure 4-11

When the *Load or Reload Linetypes* dialog box opens, you can view the entire list of available layers. Note the ACAD_ISO* files are appropriate for metric files as shown in Figure 4-12. If you only need a few linetypes, hold down the CTRL key and select from the list. You can right-click over the list and choose Select All or Clear All. After selecting the linetypes click OK to dismiss the dialog box.

Figure 4-12

When the *Select Linetype* dialog box reopens as shown in Figure 4-13, choose the linetype you need from the list as shown and click OK to dismiss the Select Linetype dialog.

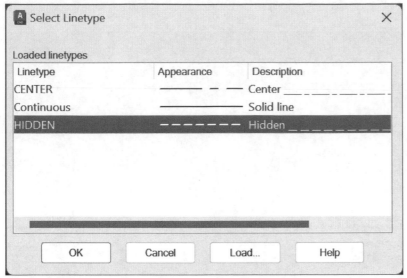

Figure 4-13

The linetype for the layer is shown in Figure 4-14.

S..	Name	On	Freeze	Lock	Plot	Color	Linetype	Lineweight
✓	0	♀	☀	🔓	🖶	■ white	Continu...	—— Default
📂	Footing	♀	☀	🔓	🖶	☐ green	HIDDEN	—— Default

Figure 4-14

Setting Layer Lineweight

Lineweight can be set in mm or inches. To view the current setting, right-click the Lineweight toggle of the Status bar and choose Lineweight Settings as shown at p1 in Figure 4-15 to open the Lineweight Settings dialog box (the toggle visibility may have to be toggled on via the Customization icon to the far right). Select the Units for Listing at p2 as shown in Figure 4-15. Click OK to dismiss the Lineweight Settings dialog box.

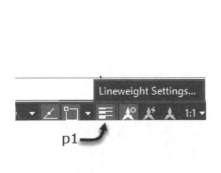

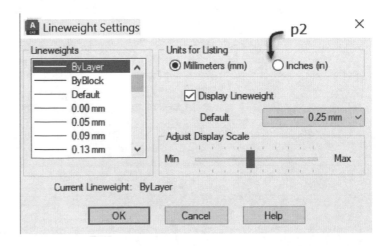

Figure 4-15

To set the lineweight for a layer, click at p1 as shown in Figure 4-16 to open the Lineweight dialog. Select a lineweight from the list and choose OK to dismiss the dialog box.

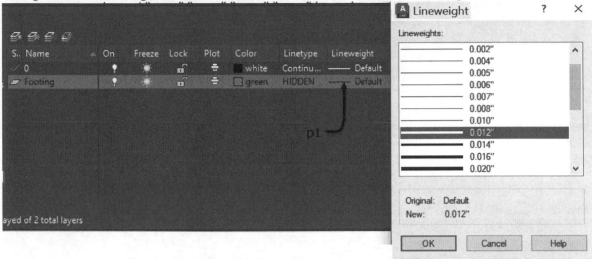

Figure 4-16

Placing Entities on Layers

To set a layer current, you can select a layer and choose the Set Current button on the toolbar, as shown in Figure 4-17. Tip: A layer can be set current by double-clicking on the layer name in the Layer Properties dialog box. A green checkmark is placed in front of the name of the current layer, as shown at p1 in Figure 4-17.

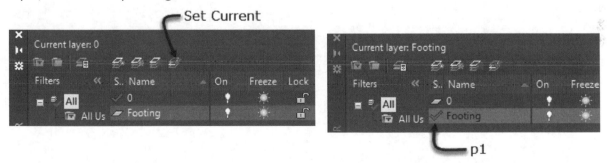

Figure 4-17

The current layer is listed in the Layer panel as shown at p1 in Figure 4-18 when no entities are selected. If an entity is selected, you can select the drop-down shown at p2 in Figure 4-18 to change the layer assignment.

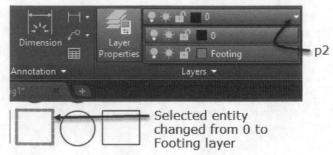

Selected entity
changed from 0 to
Footing layer

Figure 4-18

4.1 Using Properties and Quick Select

The following tutorial includes exercises using the Properties Palette and Quick Select tools.

1. Verify **Drafting** & **Annotation** is the current workspace, as shown in the Quick Access toolbar.
2. Choose Open from the Quick Access toolbar.
3. Navigate to the **AutoCAD 2024 Certified User Exercise Files \Exercise Files \Ch 4** folder. Select the **Properties**.dwg drawing file.
4. Verify **Dynamic Input** is toggled on in the Status bar. Toggle off Object Snap and toggle ON Polar.
5. The linetype of the entities shown in Figure 4-19 is the result of their layer properties.

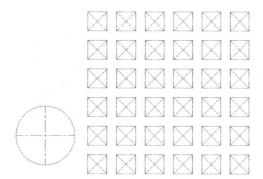

Figure 4-19

6. Choose **Select All** entities from the Utilities panel as shown at p1 in Figure 4-20. Choose Properties launching arrow shown at p2 in Figure 4-20 to display the Properties palette.
7. Choose Quick Select of the Properties palette as shown at p3 in Figure 4-20.

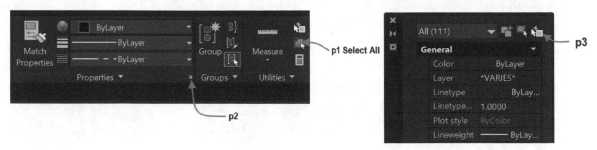

Figure 4-20

8. In the next series of steps, you will select all entities that reside on the Hidden layer and copy the content to another location.

9. Edit the *Quick Select* dialog box. In the Properties category shown in Figure 4-21 choose Layer, verify the Operator is Equals, Value is Hidden, and verify the radio button of *Include in new selection set* is toggled on in the How to apply section. Click **OK** to dismiss the Quick Select dialog box.

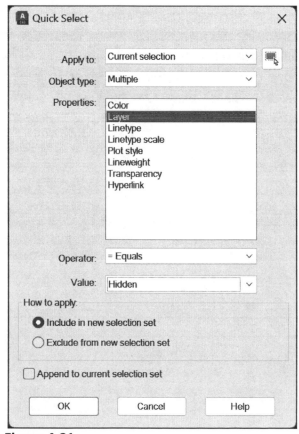

Figure 4-21

10. Retain the selection set, choose the Copy command of the Home tab, Modify panel.
Respond to the workspace prompts as follows:

Specify base point or [Displacement/mOde] <Displacement>: **9,4** (Type 9,4, press Enter to specify the base point.)

Specify second point or [Array] <use first point as displacement>: **@60,0** (Type @60,0, press Enter to specify the displacement.)

The entities of the Hidden layer copied from the coordinate 9,4 to 60,0 as shown in Figure 4-22.

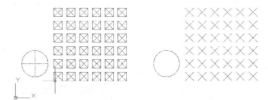

Figure 4-22

11. In this step you will use Quick Select to determine how many lines reside on the hidden layer in the drawing.

To select all entities in the drawing choose Select All entities from the Utilities panel. Choose Quick Select from the Properties palette. Edit the Quick Select dialog box as follows: Object type to **Line**, Properties to **Layer**, Value to **Hidden**, and How to apply set to *Include in new selection set* as shown in Figure 4-23.

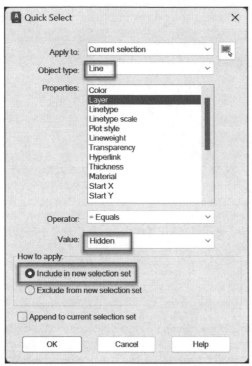

Figure 4-23

12. The Properties palette includes the number Lines on the Hidden layer as shown in Figure 4-24 at p1.

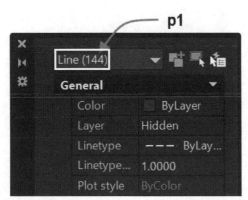

Figure 4-24

13. Using Quick Select determine how many polylines are included in the drawing. (Answer: 36)
14. Using the Properties Palette what is the circumference of the circle in the drawing? (Answer: 2' – 11 ½")
15. Using the Properties Palette, what is the coordinate of the center of the circle shown at left in the drawing? (Answer: Center X = 1'- 4 ¼" Center Y = 1' - 2 1/16")

16. Select all entities of the drawing; use Quick Select to determine how many lines in the drawing are at the angle of 322.74 degrees. (Answer: 72 lines)
17. **Save** the drawing as Properties_Your Name in your student directory and close the drawing.

4.2 Creating Layers with Linetypes

The following tutorial includes creating layers with linetype, lineweight, color properties and developing three view drawings. The tutorial assumes you have mastered the skills of developing orthographic views and the use of object snap modes.

1. Launch AutoCAD from the desktop.
2. Click New from the Start tab as shown at p1 in Figure 4-25.

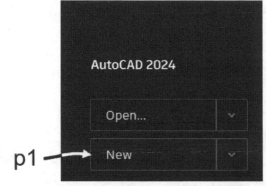

Figure 4-25

3. Verify the Drafting & Annotation workspace is current in the Quick Access toolbar shown in 4-26.

Figure 4-26

4. The Application Menu is available from all workspaces. Choose the Application Menu as shown at p1 in Figure 4-27. To set the units, choose **Drawing Utilities > Units** from the Application Menu as shown in Figure 4-27.

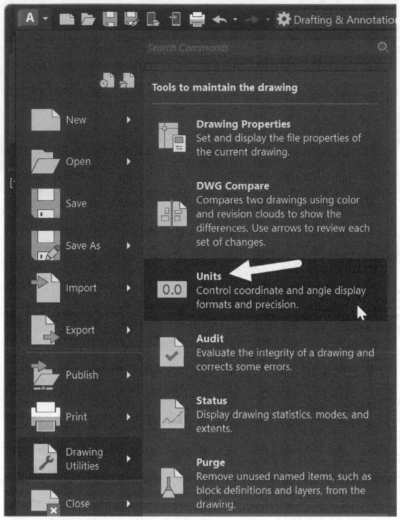

Figure 4-27

5. The Drawing Units dialog box opens, as shown in Figure 4-28. The Length type is currently set to Decimal. Click the **Length Type** dropdown shown at p1 and choose **Architectural**. Click **OK** to close the Drawing Units dialog box.

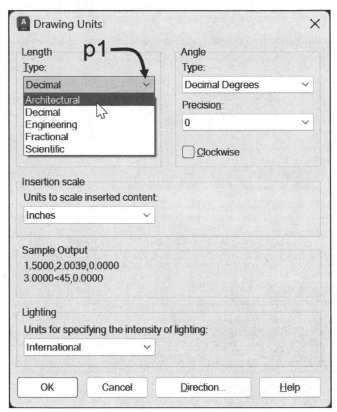

Figure 4-28

6. Choose the **Home** tab. Choose **Layer Properties** from the **Layers** panel as shown in Figure 4-29.

Figure 4-29

7. Choose **New Layer** from the Layer properties palette as shown in Figure 4-30.

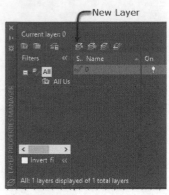

Figure 4-30

8. Overtype **Center** as the name of the new layer. Double click the **Center** name to make the new layer current. The green checkbox is displayed for the current layer as shown in Figure 4-31.

Figure 4-31

9. Choose **Customization** toggle of the Status bar and toggle ON Lineweights. Right-click the Lineweight toggle of the Status bar as shown in Figure 4-32 and choose **Lineweight Settings**.

Figure 4-32

10. Choose the Inches radio button shown at p1 in Figure 4-33 to set lineweight to inches. Choose OK to dismiss the Lineweight Settings dialog.

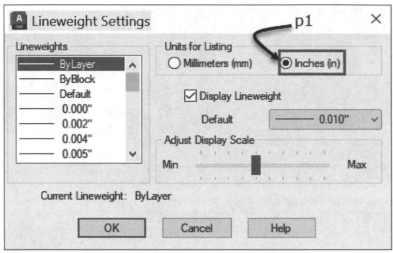

Figure 4-33

11. Return to the Layer Properties Manager palette. Click in the **Lineweight** column at p1 as shown in Figure 4-34, to open the **Lineweight** dialog box. Edit the lineweight of the **Center** layer to **.012"** as shown in Figure 4-34. Click **OK** to dismiss the dialog box.

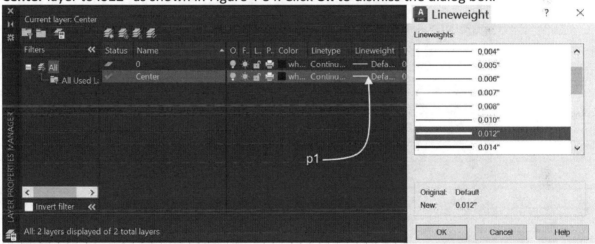

Figure 4-34

12. Click the **Color** column at p1 for the Center layer as shown in Figure 4-35 to open the Select Color dialog box. Select the **Green** color shown at p2. Click **OK** to dismiss the Select Color dialog box.

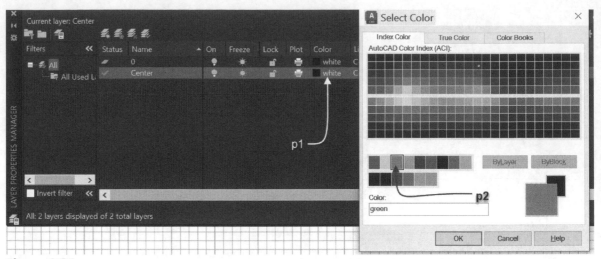

Figure 4-35

13. Click in the **Linetype** column at p1 as shown in Figure 4-36 to open the Select Linetype dialog box. Choose the **Load** button of the Select Linetype dialog box shown at p2 in Figure 4-36 to open the Load or Reload Linetypes dialog box. Scroll down the list of linetypes to select **Center** as shown at p3 in Figure 4-36. Click **OK** to all dialog boxes to assign the Center linetype to the Center layer.

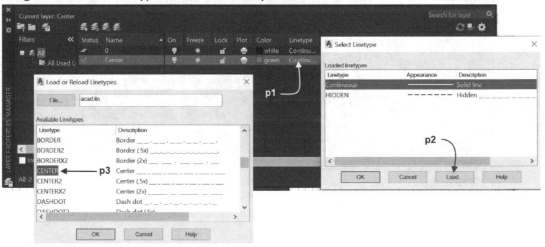

Figure 4-36

14. Continue in the Layer Properties Manager to create the following layers and properties:

Name	Color	Linetype	Line-weight
Object	Blue	Continuous	.021
Hidden	Green	Hidden	.012
Center	Green	Center	.012
Text	Green	Continuous	.012
Titleblock	Cyan	Continuous	.024
Viewport	Yellow	Continuous	.010
Dimensions	Green	Continuous	.012

Set the **Object** layer current in the Layers panel as shown at p1 in Figure 4-37.

Figure 4-37

15. Choose the **Line** command of the Draw panel. Draw three orthographic views of the Column Base shown in isometric in Figure 4-38. Note use object snap tracking presented in Chapter 3 when creating adjacent views.

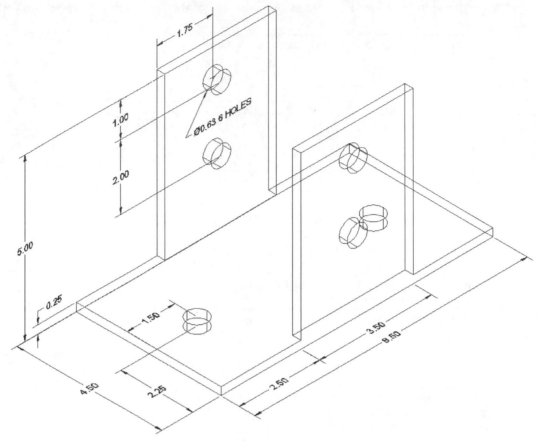

Figure 4-38

16. Choose **Layout 1** as shown at p1 in Figure 4-39 to make Layout 1 current.

Figure 4-39

17. Double-click near the hidden line, which represents the margin of the paper for the layout, as shown in Figure 4-40. The Paper Space icon should now be displayed in the lower left corner of the screen.

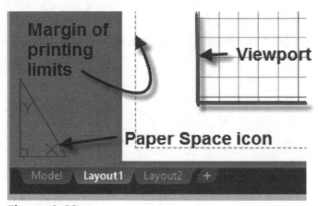

Figure 4-40

18. Select the viewport entity to display its grips as shown in Figure 4-41.

Figure 4-41

19. In this step, you will change the layer assignment of the viewport. Retain the selection of the viewport and choose the **Viewport** layer from the layer flyout in the Properties palette as shown at p1 in Figure 4-42.

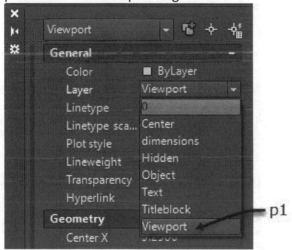

Figure 4-42

20. Toggle **ON** Lineweights in the Status bar as shown at p1 in Figure 4-43.

Figure 4-43

21. Double click inside the viewport.
    Choose **6" = 1'-0"** scale from the Viewport flyout as shown in Figure 4-44.

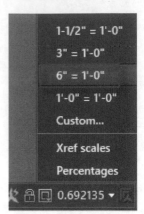

Figure 4-44

22. Due to the variety of line lengths the hidden lines do not display the gaps in the line as shown in Figure 4-45.

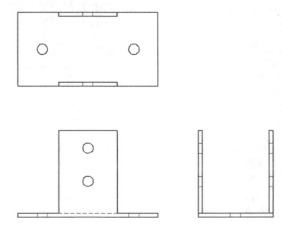

Figure 4-45

23. Type **Ltscale** in the command line and change the value to .25 to display all hidden lines shown in 4-46.

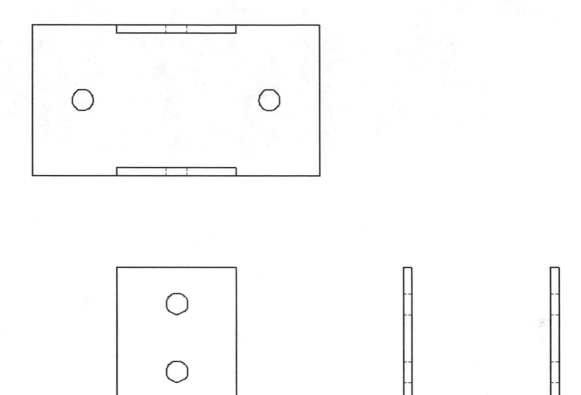

Figure 4-46

24. Set the **Center** layer current and draw a line using object tracking to track from the midpoint of the line at p1 a distance of .25 to p2 and draw a 9" line to p3 as shown in Figure 4-47.

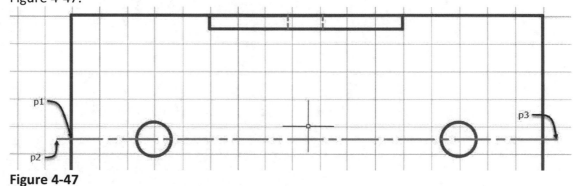

Figure 4-47

25. Choose the Layer Property Manager from the Home tab. When the Layer Property Manager opens, notice additional columns have been added with VP prefixing the name. You can choose the **VP Freeze** toggle shown at p1 in Figure 4-48 to freeze the

center layer in only this viewport. The center line will be visible in the Model tab and Layout 2 tabs.

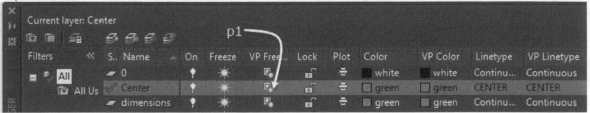

Figure 4-48

26. Click **Save** on the **Quick Access toolbar**. Save the file as 4.2 Layers and Linetypes Your Name to your student folder. Choose Close from the Application Menu to close the file.

Chapter 5
Modifying Objects

This chapter includes a review of the commands used to modify the entities of the drawing. The commands such as offset, trim, extend, and move are presented with emphasis on noun verb selection and workspace prompts using dynamic input. The fillet chamfer and blend commands are presented for editing lines, polylines, polygons, and rectangles. Finally, the use of grips for lengthen, stretch, move, copy, mirror, and rotate are included. There are eight tutorials, and a Grips Quiz included to help you prepare for the test. The content of this chapter is very important for test preparation. Since the test is timed, it would be helpful to repeat the tutorials to decrease the time for editing and develop confidence.

The student will be able to:

1. Use the Blend command to create a spline between entities
2. Create a gap in entities using the Break command
3. Edit the length of entities using the Trim and Extend commands
4. Modify the intersection of entities to create fillets and chamfers
5. Connect two or more collinear entities with the Join command
6. Use the Offset command to copy entities parallel to an entity
7. Copy and move entities using the Copy and Move commands
8. Copy or move entities about an axis line using the Mirror command
9. Change the length and width of closed shapes using the Scale command
10. Shorten or lengthen entities using the Stretch command
11. Rotate entities about a base point at a specified angle
12. Erase entities from the drawing

Copy / Move

The modify objects tools are located on the Modify panel and the panel extension as shown in Figure 5-1. To extend the panel click the flyout shown at p1 to display the commands as shown at p2 in Figure 5-1.

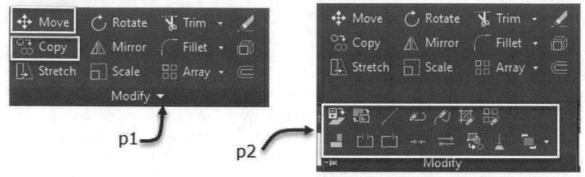

Figure 5-1

The Move and Copy commands are similar in workspace prompts most often included in a test session. The copy command includes the copy multiple mode which by default you can enter additional distances to copy the selected entity from the base point. The use of the Copy command to copy the circle shown at p1 in Figure 5-2 to a point .75 to the right follows: Note you can open the **Move demo** from the **AutoCAD 2024 Certified User Exercises Files \ Exercise Files \ Certification Demo Files \ Ch 5** folder to try this technique as you read the steps.

Toggle on Dynamic Input and Center Object snap mode in the Status bar.
Select the circle shown at p1 in Figure 5-2 and choose **Copy** from the Modify panel.
Command: Copy 1 found
Current settings: Copy mode = Multiple
Specify base point or [Displacement/mOde] <Displacement>: (Select the center of the circle using the **Center** object snap mode; move the cursor to the right as shown at p2 in Figure 5-2).
Specify second point or [Array] <use first point as displacement>: (Overtype .**75** press Tab, type **0** in the polar angle field, press ENTER to specify the second point as shown at p3 in Figure 5-2.)
Specify second point or [Array/Exit/Undo] <Exit>: **ENTER** (Press Enter to end the command. Circle copied as shown at p4 in Figure 5-2.)

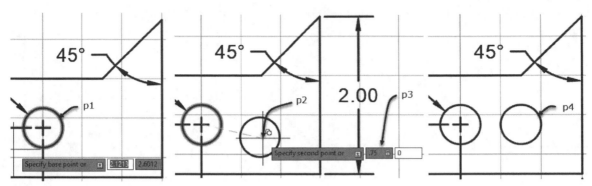

Figure 5-2

Move

The technique to use the Move command to move a circle shown at p1 in Figure 5-3 to a point .75 to the right follows. Note you can open the **Move demo** from the **AutoCAD 2024 Certified User Exercises Files \ Exercise Files \ Certification Demo Files \ Ch 5** folder to try this technique as you read the steps.

Toggle on Dynamic Input and Center Object snap mode in the Status bar.
Select the circle shown at p1 in Figure 5-3; choose **Move** from the Modify panel in Home tab shown in Figure 5-1.
Command: _move 1 found
Specify base point or [Displacement] <Displacement>: (Select the center of the circle using the **Center** object snap mode. Move the cursor to the right as shown at p2 in Figure 5-3).
Specify second point or <use first point as displacement>: @.75<0 (Overtype **.75** in the distance field, press Tab, type **0** in the polar angle field, and press **ENTER** to specify the second point as shown at p3 in Figure 5-3.)
The circle is moved as shown at p4 in Figure 5-3.

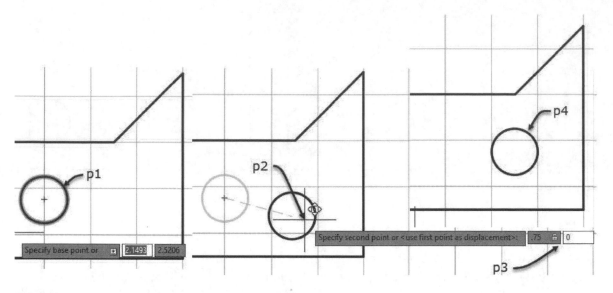

Figure 5-3

Erase

The Erase command is located on the Modify panel as shown in Figure 5-4. The technique to use the command is very simple—you just select the entities you need erased and choose **Erase** on the Modify panel or Delete on the keyboard.

Figure 5-4

Extend and Trim

The Extend command is located on the Modify panel in the Home tab. Click the flyout as shown at p1 in Figure 5-5 to display the Trim and Extend commands. The selected command stays on top when the menu collapses.

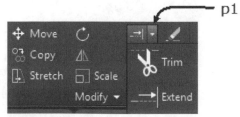

Figure 5-5

The technique to use the Extend command is shown in the following workspace prompts.

Note you can open the **Extend demo** from the **AutoCAD 2024 Certified User Exercises Files \ Exercise Files \ Certification Demo Files \ Ch 5** folder to try this technique as you read the following steps.

The Extend command can be used to extend lines, polylines, arcs, elliptical arc and splines to other entities. The default settings for the Extend Trim commands in prior releases required you to select the entities that will serve as a boundary for entities when extended. As shown in Figure 5-6, line p3 will not serve as a boundary for line p4 since p4 if projected will not butt to p3, whereas line p1 will serve as a boundary for line p2 as shown in Figure 5-6.

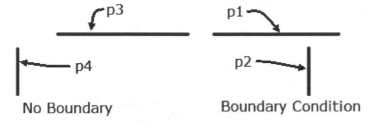

Figure 5-6

However current settings in AutoCAD 2024 presets the Mode to Quick as shown in the following command prompts:
Command: _extend
Current settings: Projection=UCS, Edge=None, Mode=Quick
Select object to extend or shift-select to trim or
 [Boundary edges/Crossing/mOde/Project]:

The Quick mode presets all entities of the drawing to act as a boundary without requiring you to select an entity to serve as a boundary. The Quick mode reduces your operations and should allow you to accomplish your trim quicker.

Tip: You can change the Mode from Quick to Standard which will require you to select specific entities as boundaries. The Standard mode allows you to change Edge to Extend which allows line p3 to serve as a boundary for line p4 as shown in Figure 5-6. **Caution if you follow the procedure shown below you must repeat the procedure to reset the Mode to Quick and Edge to No Extend to regain Extend and Trim default operations for AutoCAD 2024 in all future drawings of the login.**

Current settings: Projection=UCS, Edge=None, Mode=Quick
Select object to extend or shift-select to trim or
 [Boundary edges/Crossing/mOde/Project]: **O**
Enter an extend mode option [Quick/Standard] <Quick>: **S**
Select object to extend or shift-select to trim or
 [Boundary edges/Fence/Crossing/mOde/Project/Edge]: **E**
Enter an implied edge extension mode [Extend/No extend] <No extend>: **E**
Select object to extend or shift-select to trim or

[Boundary edges/Fence/Crossing/mOde/Project/Edge/Undo]:

The technique to use the Extend command is shown in the following workspace prompts. Note that you can open the **Extend demo.dwg** drawing file from the **AutoCAD 2024 Certified User Exercises Files \ Exercise Files \ Certification Demo Files \ Ch 5** folder to try this technique as you read the following steps.

Choose the **Extend** command from the Modify panel as shown at right in Figure 5-7. (Notice the dynamic input prompt to select objects to extend includes the option to hold the shift key down to switch to trim when you select objects.)
Specify first fence point or shift select to trim: (Select a point near p1 in Figure 5-7; do not release the mouse button.)
Specify next fence point or: (Move the mouse to a point near p2 in Figure 5-7, then release the left mouse button.)
Select object to extend or shift-select to trim or: (Press Enter to end the command.)

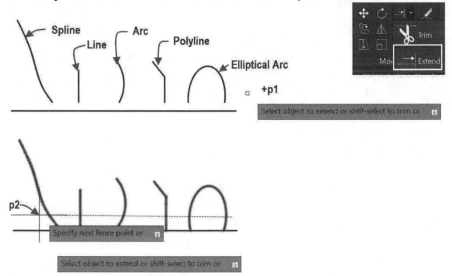

Figure 5-7

Trim

The Trim command can be chosen from the Home tab, Modify panel by choosing the scissor icon of the Trim Extend flyout as shown in Figure 5-8. The Trim command includes the Quick Mode, and you are not prompted to select a cutting edge since all objects will serve as cutting edges by default. Notice that the prompt to select objects to trim includes the option to hold the shift key down and switch to extend when you select objects to trim. The technique to use the Trim command is shown in the following workspace prompts, which include the command line content and Figure 5-8 includes dynamic prompts. Press F2 function key to display current settings which are not displayed in the dynamic prompts.

Note that you can open the **Trim demo.dwg** drawing file from the **AutoCAD 2024 Certified User Exercises Files \ Exercise Files \ Certification Demo Files \ Ch 5** folder to try this technique as you read the following steps.

Choose the **Trim** command from the Modify panel as shown at left in Figure 5-8.
Command: _trim
Current settings: Projection=UCS, Edge=None, Mode=Quick
Select object to trim or shift-select to extend or
 [cuTting edges/Crossing/mOde/Project/eRase]: Select the line as shown at p1 in Figure 5-8.
Select object to trim or shift-select to extend or
 [cuTting edges/Crossing/mOde/Project/eRase/Undo]: Select the line as shown at p2 in Figure 5-8.
Select object to trim or shift-select to extend or [cuTting
edges/Crossing/mOde/Project/eRase/Undo]: Press **ENTER** to end the command.

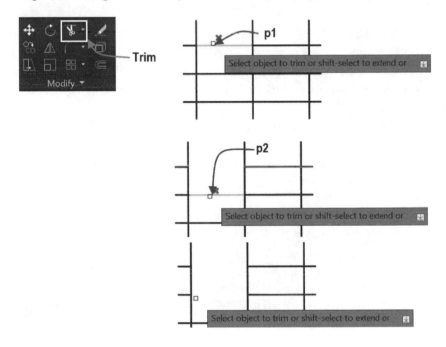

Figure 5-8

Fillet / Chamfer

The Fillet, Chamfer, and Blend Curves commands are accessed from the flyout shown at p1 of the Modify Panel in the Home tab in Figure 5-9. Note that you can choose the Polyline option when prompted to select the first line to select a polyline and apply the fillet to all corners or vertices of the rectangle, polygon, and polylines. The Fillet command can create a corner or a radius intersection between two lines as shown in Figure 5-9. If lines are parallel, as shown at C in Figure 5-9, the Fillet command will determine the radius as necessary to create a slot.

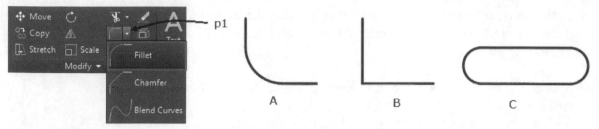

Figure 5-9

The technique to use the Fillet command is shown in the following workspace prompts and Figure 5-10. Open the **Fillet chamfer demo.dwg** drawing file from the **AutoCAD 2024 Certified User Exercises Files \ Exercise Files \ Certification Demo Files \ Ch 5** folder to try this technique as you read the steps.

Choose the **Fillet** command from the flyout as shown at p1 in Figure 5-9.
FILLET
Current settings: Mode = TRIM, Radius = 0.0000
Select first object or: *(Select the line shown at **p1** in Figure 5-10.)*
Select second object or shift-select to apply corner or: **R** *(Press the down arrow on the keyboard to display the drop-down options menu as shown at **p2** in Figure 5-10. Press the down arrow on the keyboard to choose the radius option.)*
Specify fillet radius: **.375** *(Type **.375** in the radius field as shown at **p3** in Figure 5-10 and press **Enter**.)*
Select second object or shift-select to apply corner or: *(Select the line as shown at **p4** in Figure 5-10. The fillet is created as shown at **p5** in Figure 5-10.)*

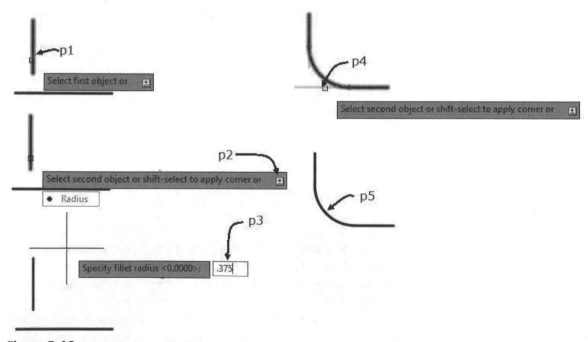

Figure 5-10

Chamfer

The Chamfer command modifies the intersection by adding an angular flat as shown in Figure 5-11. The chamfer can be applied by specifying distances or the intersection angle. Note, you can choose the Polyline option of the command when prompted to *Select first line* to apply the chamfer to all corners or vertices of the rectangle, polygon and polylines.

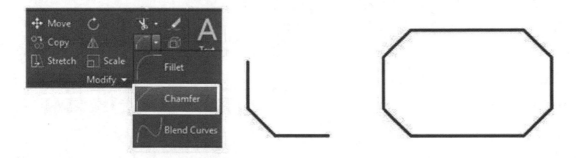

Figure 5-11

The technique to use the Chamfer command is shown in the following workspace prompts and Figure 5-12. Open the **Fillet chamfer demo.dwg** drawing file from the **AutoCAD 2024 Certified User Exercises Files \ Exercise Files \ Certification Demo Files \ Ch 5** folder to try this technique as you read the steps.

Choose the **Chamfer** command from the Fillet Chamfer Blend flyout shown in Figure 5-11.
CHAMFER
(TRIM mode) Current chamfer Dist1 = 0.0000, Dist2 = 0.0000
Select first line or: D (Press the down arrow on the keyboard to display the drop-down options menu as shown at p1. Press the down arrow on the keyboard to choose the Distance option.)
Specify first chamfer distance <0.0000>: **.25** *(Type .25 in the Distance field as shown at p2 in Figure 5-12 and press Enter.)*
Specify second chamfer distance <0.2500>: **.5** *(Type .25 in the Distance field as shown at p3 in Figure 5-12 and press Enter.)*
Select first line or: *(Select the vertical line as shown at p4 in Figure 5-12.)*
Select second line or shift-select to apply corner or: *(Select the line as shown at p5 in Figure 5-12.)*
The completed chamfer is shown at p6 in Figure 5-12.

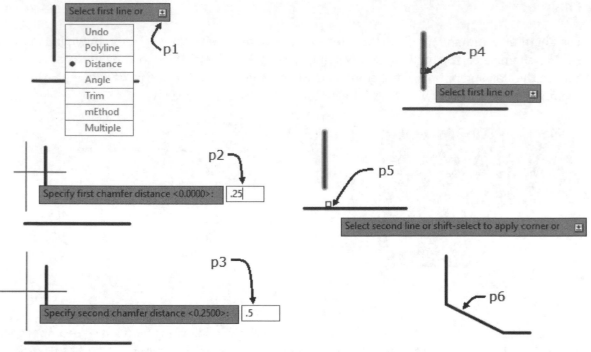

Figure 5-12

Blend Curve

The Blend Curves command creates a spline that connects to the ends of lines or arcs as shown in Figure 5-13. The Blend Curves command, as shown at p1 in Figure 5-13, is accessed from the Fillet Chamfer Blend flyout.

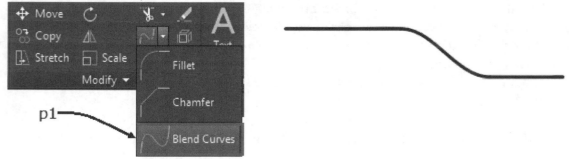

Figure 5-13

The technique to use the Blend command is shown in the following workspace prompts and Figure 5-14. The Open **Fillet chamfer demo.dwg** drawing file from the **AutoCAD 2024 Certified User Exercises Files \ Exercise Files \ Certification Demo Files \ Ch 5** folder to try this technique as you read the following steps.

Choose the **Blend** command from the Fillet / Chamfer / Blend flyout shown at p1 in Figure 5-13. Command: _BLEND

Continuity = Tangent

Select first object or: (Select the line at p1 as shown in Figure 5-14.)
Select second object: (Select the line at p2 as shown in Figure 5-14.)
The blended curve is created as shown at p3 in Figure 5-14.

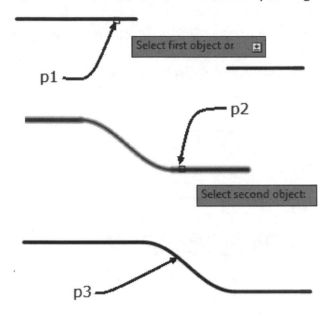

Figure 5-14

Offset

The offset command copies an entity such as a circle, line, polyline, arc, or polygon parallel to the specified distance. The Offset command is on the Modify panel in the Home tab, as shown at p1 in Figure 5-15. The command includes a Multiple option, which allows you to make multiple copies of the selected object. The technique to use the Offset command is shown in the following workspace prompts and Figure 5-15. The Multiple option of the command will be used in this example to create four copies of a line ½" apart. Note that you can open the **Offset demo.dwg** drawing file from the **AutoCAD 2024 Certified User Exercises Files \ Exercise Files \ Certification Demo Files \ Ch 5** folder to try this technique as you read the steps.

Select the line as shown at p2 in Figure 5-15 to apply the offset command.
Choose the Offset command as shown at p1 in Figure 5-15.
_offset
Current settings: Erase source=No Layer=Source OFFSETGAPTYPE=0
Specify offset distance or [Through/Erase/Layer] <Through>: .5 (Type **.5** and press **Enter** in the workspace prompt as shown at p3 in Figure 5-15.)
Specify point on side to offset or [Exit/Multiple/Undo] <Exit>: **M** (Press the down arrow on the keyboard to display the drop-down options menu as shown at p4 in Figure 5-15. Press the down arrow on the keyboard to choose the **Multiple** option.)

Specify point on side to offset or [Exit/Undo] <next object>: (Move the cursor near p5, left click to create a copy.)
Specify point on side to offset or [Exit/Undo] <next object>: (Move the cursor near p5, left click to create a copy.)
Specify point on side to offset or [Exit/Undo] <next object>:(Move the cursor near p5, left click to create a copy.)
Specify point on side to offset or [Exit/Undo] <next object>:(Move the cursor near p5, left click to create a copy.)
Select object to offset or [Exit/Undo] <Exit>: (Press **Enter** to end the command.)

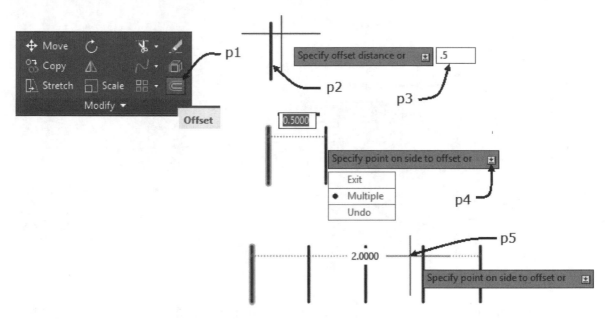

Figure 5-15

Mirror

The Mirror command can be used to copy entities from half of a symmetrical layout to complete the other half. Access the mirror command from the Modify panel in the Home tab as shown at p3 in Figure 5-16. The technique to use the Mirror command is shown in the following workspace prompts and Figure 5-16. Note you can open the **Mirror demo.dwg** drawing file from the **AutoCAD 2024 Certified User Exercises Files \ Exercise Files \ Certification Demo Files \ Ch 5** folder to try this technique as you read the following steps.

Select the three circles and slot by creating a rectangular window selection from p1 to p2 as shown in Figure 5-16.
Select the Mirror command from the Modify panel as shown at p3 in Figure 5-16.

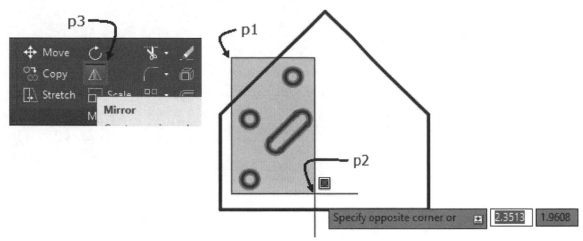

Figure 5-16

Command: _mirror 7 found
Specify first point of mirror line: (Move the cursor to p1 as shown in Figure 5-17; left click when the Endpoint object snap marker is displayed.)
Specify second point of mirror line: **@2<270** (Move the cursor down, type **2**, press **Tab** key and type **270** to specify the polar angle at p2 as shown in Figure 5-17.)
Erase source objects? [Yes/No] <No>:(Press **Enter** to retain the objects.)
The completed mirror design is shown in Figure 5-18.

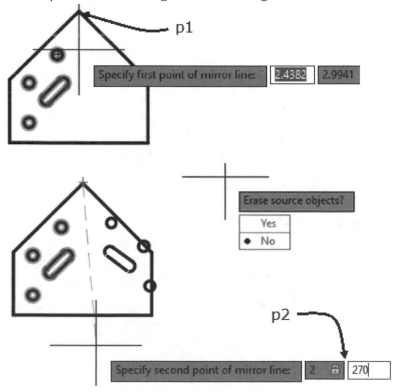

Figure 5-17

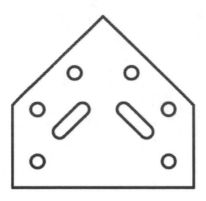

Figure 5-18

Stretch

The Stretch command can be used to move and stretch entities from part of a design to create a shorter or longer part. Access the Stretch command from the Modify panel in the Home tab as shown at p1 in Figure 5-19. When using the stretch command, you must use a crossing polygon or a crossing window when selecting the objects for edit. The technique to use the Stretch command is shown in the following workspace prompts and Figure 5-19. Note you can open the **Stretch demo.dwg** drawing file from the **AutoCAD 2024 Certified User Exercises Files \ Exercise Files \ Certification Demo Files \ Ch 5** folder to try this technique as you read the steps.

Select the three circles and slot by creating a rectangular crossing selection from p2 to p3 as shown in Figure 5-19.
Select the **Stretch** command from the Modify panel as shown at p1 in Figure 5-19.

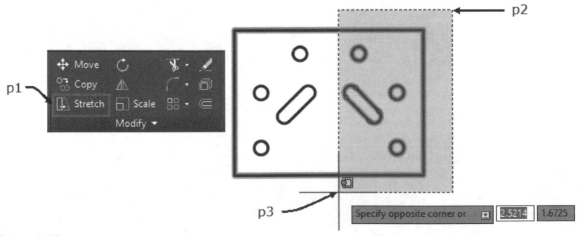

Figure 5-19

Command: _stretch
Stretching selected objects by last window...8 found

Specify base point or: (Click near p1 as shown in Figure 5-20 to specify the basepoint.)
Specify second point or <use first point as displacement>: @1<0 (Move the cursor right, type **1** in the distance field, press the **Tab** key to move the focus to the polar angle, type **0**, press **ENTER** as shown at p2 in Figure 5-20.)
The stretched part is shown in Figure 5-21.

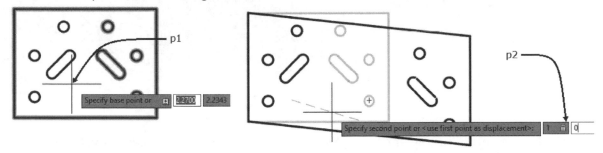

Figure 5-20

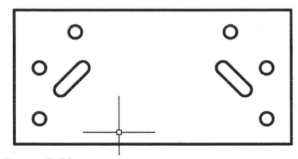

Figure 5-21

Scale

The scale command can be used to increase or decrease the size of a part. To increase the size, enter a whole number, and to decrease the size, enter a fraction in the scale factor field. Access the Scale command from the Modify panel in the Home tab, as shown at p1 in Figure 5-22. The technique to use the Scale command is shown in the following workspace prompts.

Note you can open the **Scale demo.dwg** drawing file from the **AutoCAD 2024 Certified User Exercise Files \ Exercise Files \ Certification Demo Files \ Ch 5** folder to try this technique as you read the steps.

Select the part shown by creating a rectangular window selection from p2 to p3 as shown in Figure 5-22.
Choose the **Scale** command from the Modify panel as shown at p1 in Figure 5-22.

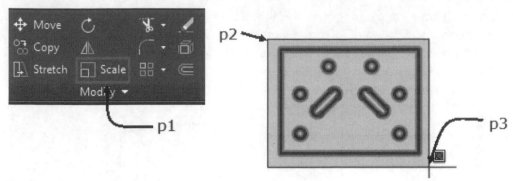

Figure 5-22

Command: _scale 15 found
Specify base point: (Click the endpoint, the line at p1 in Figure 5-23, to specify the basepoint.)
Specify scale factor or [Copy/Reference]: R (Press the **down arrow** on the keyboard to display the command options drop-down menu as shown at p2 in Figure 5-23. Press the down arrow twice to select the **Reference** option.)
Specify reference length <1.7273>: (Select a point using the endpoint object snap at p3 in Figure 5-23.)
Specify second point: (Select a point using the endpoint object snap at p4 in Figure 5-23.)
Specify new length or [Points] <2.0000>: 2.5 (Type **2.5** to specify the new length at p1 in Figure 5-24. Scaled part is shown at right in Figure 5-24.)

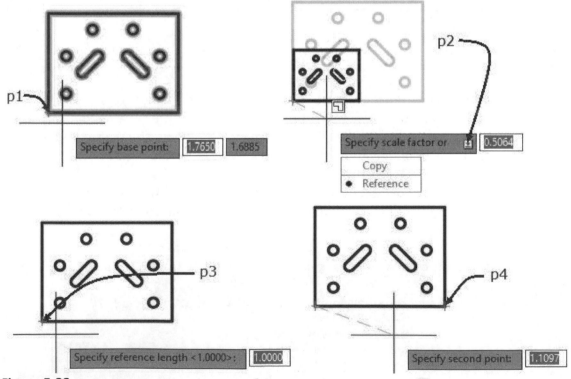

Figure 5-23

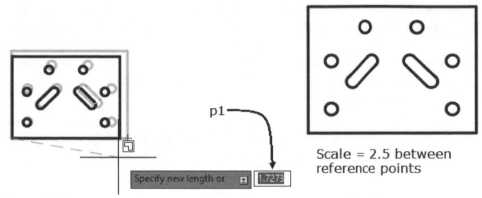

Scale = 2.5 between
reference points

Figure 5-24

Rotate

The Rotate command is used to rotate entities at a specified angle about the basepoint.
Positive angles are measured in a **counterclockwise** direction from the 3:00 o'clock reference
point. The rotate command also includes a reference option to specify rotation to match the
rotation of existing entities.

Access the Rotate command from the Modify panel in the Home tab, as shown at p1 in Figure
5-25. The technique to use the Rotate command is shown in the following workspace prompts
using the Reference option.

Note you can open the **Rotate demo** from the **AutoCAD 2024 Certified User Exercise Files **
Exercise Files \ Certification Demo Files \ Ch 5 folder to try this technique as you read the
steps.

Toggle on only **Geometric Center** and **Center** object snaps in the Status bar.
Select the part shown by creating a rectangular window selection from p2 to p3 as shown in
Figure 5-25.
Select the Rotate command from the Modify panel as shown at p1 in Figure 5-25.

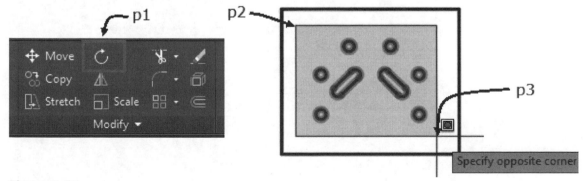

Figure 5-25

Command: _rotate
Current positive angle in UCS: ANGDIR=counterclockwise ANGBASE=0
Select objects: Specify opposite corner: 15 found

Specify base point: (Move the cursor over the perimeter of the rectangle and click when the Geometric Center object snap marker is displayed at p1 in Figure 5-26 to specify the basepoint.)

Specify rotation angle or [Copy/Reference] <0>: R (Press the down arrow on the keyboard to display command options drop-down menu as shown at p2 in Figure 5-26. Press the down arrow twice to select the Reference option.)

Specify the reference angle <0>: (Move the cursor over the circle shown at p3 in Figure 5-26; click when the center object snap marker is displayed to select the center.)
Specify second point: (Move the cursor over the circle shown at p4 in Figure 5-26; click when the center object snap marker is displayed to select the center.)
Specify the new angle or [Points] <0>: 0 (Type **0** Press **Enter** at p1 to specify the angle as shown in Figure 5-27. The selected entities are shown rotated at p2 in Figure 5-27.)

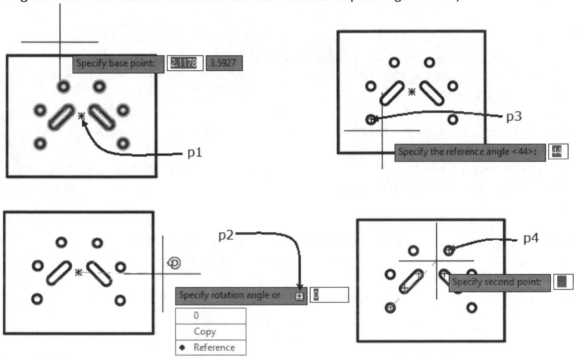

Figure 5-26

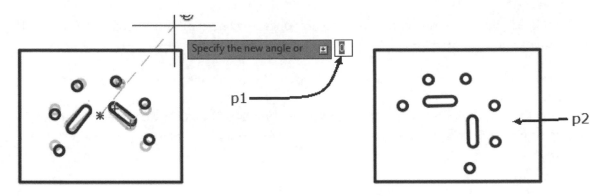

Figure 5-27

Break / Join

Access the Break and Join commands from the extended Modify panel in the Home tab, as shown in Figure 5-28. The Break at Point command allows you to break a line, arc, or an open polyline at a specified point. Note that the Break at Point command can be repeated by pressing Enter upon selecting a point.

Open the **Break Join demo.dwg** drawing file from the **AutoCAD 2024 Certified User Exercises Files \ Exercise Files \ Certification Demo Files \ Ch 5** to follow the following steps. The technique to use the Break at Point command is shown in the following workspace prompt and in Figure 5-28.

Choose the Break at Point command as shown in Figure 5-28.
Select objects: (Select the polyline at p1 as shown in Figure p1 as shown in Figure 5-28.)
Specify break point: (Select a location on the polyline at p2 as shown in Figure 5-28.)
(Select a segment of the polyline to view the break and the grips of the polyline as shown at p3 in Figure 5-28.)
Note the single polyline has been changed to two polyline segments.

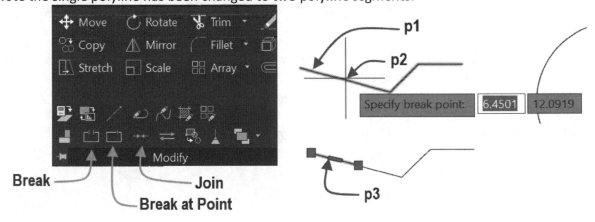

Figure 5-28

The Break command creates a gap in lines between two points selected. The command is useful in editing lines for clarity in electrical and plumbing line diagrams. The first point selected starts the break; as you move the cursor from the first point the line slides back to create the gap. The point location can be selected on or off the line.

The technique to use the Break command is shown in the following workspace prompts and Figure 5-29.
Note you can open the **Break Join demo** from the **AutoCAD 2024 Certified User Exercises Files \ Exercise Files \ Certification Demo Files \ Ch 5** folder to try this technique as you read the steps.

Choose the Break command as shown in Figure 5-28.
Select object: (Select the line at p1 shown in Figure 5-29 to start the break.)
Specify second break point or [First point]: (Select the line at p2 shown in Figure 5-29 to create the gap as shown at p3.)

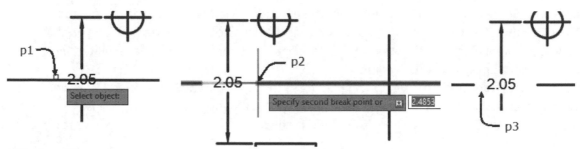

Figure 5-29

The Join command is used to join collinear line segments that may have been created by break tools. Open the **Break Join Demo.dwg** drawing file located in the **AutoCAD 2024 Certified User Exercises Files \ Exercise Files \ Certification Demo Files \ Ch 5** folder to use the Join command to join the line segments as described in the following steps.

Choose the Join command from the extended Modify panel, as shown in Figure 5-28.

Command: _join
Select source object or multiple objects to join at once: (Use the crossing window selection and click at **p1** to move the cursor left as shown in Figure 5-30.)
Specify opposite corner: 4 found (Click at **p2** to include all line segments as shown in Figure 5-30.)
Select objects to join: Press **Enter** to end the command. All four lines are joined into one line as shown at p3 in Figure 5-30.
4 lines joined into 1 line

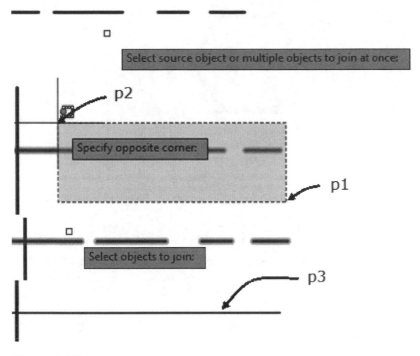

Figure 5-30

Grips

When you select an entity such as a line, polyline, polygon, or circle grip, they are displayed in blue color at such points as the center, endpoints, and midpoints as shown in Figure 5-31. Note you can open the **Grips demo.dwg** drawing file from the **AutoCAD 2024 Certified User Exercises Files \ Exercise Files \ Certification Demo Files \ Ch 5** folder to try this technique as you read.

When you hover over a grip, the color changes to pink, and if you click the grip, the color changes to red.

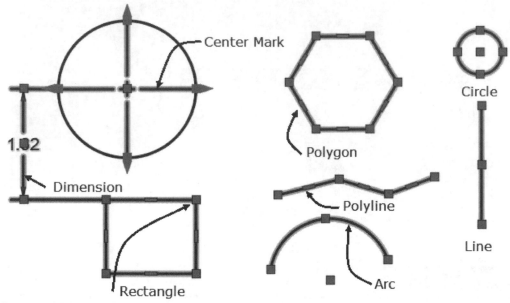

Figure 5-31

When a grip is red it is selected and a menu for editing the grip and dynamic dimensions is displayed in the workspace as shown in Figure 5-32. The contextual menu for a selected grip is shown at right in Figure 5-32. When a grip is selected press the space bar to toggle through edit options of stretch, move, rotate, scale, mirror. The grip located at the midpoint of a line allows you to change the location of the line. The grips at endpoint allow you to change the location of the end and edit using the contextual menu.

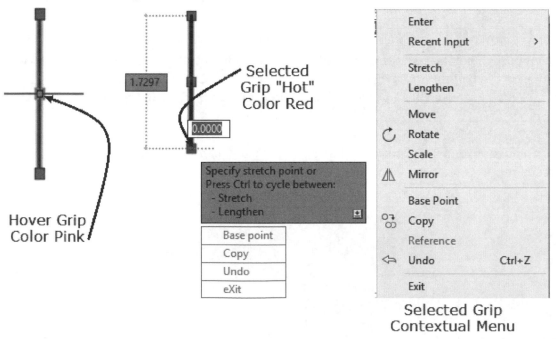

Figure 5-32

To select multiple grips, hold the Shift key down before starting the selection of grips. To deselect a grip, continue to hold down the Shift key and reselect a selected grip. To stop editing with a selected grip, press the Escape key, and to clear the display of grips, press the Escape key. When editing a grip, the grip will snap to other displayed grips without regard to the object snap settings.

A unique feature of grips is the ability to rotate copy a line the angular measure relative to an existing line. This feature is very helpful for developing property boundaries from survey data. The technique to use grips to rotate copy a line 120 degrees is shown in the following workspace prompts and Figure 5-33.

Note that you can open the **Grips demo.dwg** drawing file from the **AutoCAD 2024 Certified User Exercises Files \ Exercise Files \ Certification Demo Files \ Ch 5** folder to try this technique as you read the steps.

Select the line to display its grips as shown at p1 in Figure 5-33.
Click the grip shown at p2 in Figure 5-33; the selected grip displays in red color. Press the spacebar twice to toggle to Rotation and press the down arrow on the keyboard to display the menu as shown at p3 in Figure 5-33. Press the down arrow on the keyboard twice to toggle to Copy. Press Enter to select Copy.
Overtype **-120** in the angle field as shown at p4 in Figure 5-33. Press Escape twice to end the edit and the display of grips.
The rotated line is copied 120 degrees as shown at p5 in Figure 5-33.

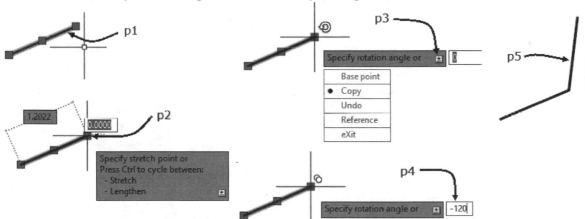

Figure 5-33

5.1 Creating Basic Fillets and Chamfers

The following tutorial includes creating fillets and chamfers of existing lines.

1. AutoCAD 2024 from the Desktop shortcut.
2. Choose **Open** from the Quick Access toolbar.

3. Navigate to the **AutoCAD 2024 Certified User Exercises Files \ Exercise Files \ Ch 5** folder and choose **Fillet 1.dwg**.

4. Save the drawing in your student folder.

5. Toggle ON Polar and Dynamic Input in the Status bar as shown in Figure 5-34.

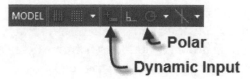

Figure 5-34

6. In this step you will create a fillet connecting the lines shown at A in the Fillet drawing. Choose **Fillet** from the Modify panel. Respond to the workspace prompts as follows:
Select first object or: (Select the line at p1 shown in Figure 5-35.)
Select second object or shift-select to apply corner or [Radius]: (Select the line at p2 as shown in Figure 5-35.)

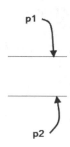

Figure 5-35

7. Repeat the Fillet command to fillet the lines shown at left in Figure 5-36 to create a slot.

Figure 5-36

8. The completed fillets are shown in Figure 5-37.

Figure 5-37

9. In the next steps you will apply the fillet to a polyline shown at B in the Fillet1 drawing. Choose the **Fillet** command from the Modify panel. Respond to the Workspace prompts as follows:

    Current settings: Mode = TRIM, Radius = 0.0000

    Select first object or [Undo/Polyline/Radius/Trim/Multiple]: P (Press the down arrow key to display the command options. Continue to press the down arrow key to toggle to the **Polyline** option; press **Enter**.)

    Select 2D polyline or [Radius]: r (Press the down arrow key to display the command options. Continue to press the down arrow key to toggle to the **Radius** option; press **Enter**.)

    Specify fillet radius <0.0000>: **.25** (Type **0.25** to specify the radius.)

    Select 2D polyline or [Radius]: (Select the Polyline at p1 as shown in Figure 5-38.)

    4 lines were filleted.

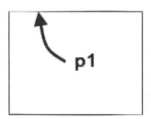

Figure 5-38

The fillet is applied to rectangle as shown in Figure 5-39.

Figure 5-39

10. In this step you will create a fillet connecting the lines shown at C in the Fillet1 drawing. Select the Fillet command of the Modify panel. Respond to the Workspace prompts as follows:

Command: fillet

Current settings: Mode = TRIM, Radius = 0.2500

Select first object or [Undo/Polyline/Radius/Trim/Multiple]: R *(Press the down arrow key to display the command options. Continue to press the down arrow key to toggle to the **Radius** option; press **ENTER**.)*

Specify fillet radius <0.0000>: **.5** *(Type .5 to set the radius.)*

Select first object or [Undo/Polyline/Radius/Trim/Multiple]: *(Select line at p1 as shown in Figure 5-40.)*

Select second object or shift-select to apply corner or [Radius]: *(Select the line at p2 as shown in Figure 5-40.)*

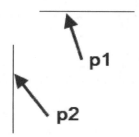

Figure 5-40

11. In the next step you will create a chamfer connecting the lines at D in the Fillet 1 drawing using the distance of .5.

12. Choose **Chamfer** from the Modify panel as shown in Figure 5-41.

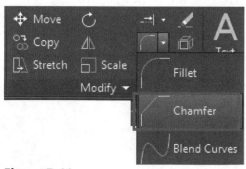

Figure 5-41

Respond to the workspace prompts as follows:

Command:_CHAMFER

(Trim mode) Current chamfer Dist1 = 0.0000, Dist2 = 0.0000

Select first object or [Undo/Polyline/Distance/Angle/Trim/mEthod/Multiple]: d (Press the down arrow key to display the command options. Continue to press the down arrow key to toggle to the **Distance** option; press **Enter**.)

Specify first chamfer distance <0.0000>: **.5** (Enter **.5** chamfer distance for the first line.)

Specify second chamfer distance <0.5000>: (Press **ENTER** to assign .5 as the chamfer distance for the second line.)

Select first object or [Undo/Polyline/Distance/Angle/Trim/mEthod/Multiple]: (Choose the line at **p1** as shown in Figure 5-42.)

Select second line or shift-select to apply corner or [Distance/Angle/Method]: (Choose the line at **p2** as shown in Figure 5-42.)

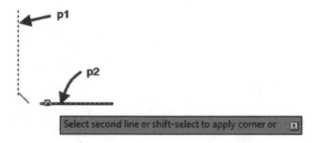

Figure 5-42

13. In the next step you will create connecting the lines shown at E in the Fillet1 drawing using the distance and angle option.

14. Choose Chamfer from the Modify panel. Respond to the workspace prompts as follows:
    Command: _chamfer
    (Trim mode) Current chamfer Dist1 = 0.5000, Dist2 = 0.5000
    Select first line or [Undo/Polyline/Distance/Angle/Trim/mEthod/Multiple]: a (Press the down arrow key to display the command options. Continue to press the down arrow key to toggle to the **Angle** option; press **Enter**.)
    Specify chamfer length on the first line <0.5000>: **1** Enter (Type **1**, press **Enter** to specify the distance.)
    Specify chamfer angle from the first line <0>: **30** Enter *(Type **30**, press **Enter** to specify the angle.)*
    Select first line or [Undo/Polyline/Distance/Angle/Trim/mEthod/Multiple]: *(Select the line at p1 as shown in Figure 5-43.)*
    Select second line or shift-select to apply corner or [Distance/Angle/Method]: *(Select the line at p2 as shown in Figure 5-43.)*

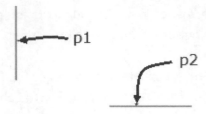

Figure 5-43

15. The chamfer created as shown in Figure 5-44.

Figure 5-44

16. Fillet each of the lines shown at **F** in the Fillet1 drawing and Figure 5-45 to create a slotted hole.

F

Figure 5-45

17. Save the drawing as Fillets_Your Name to your student folder and close the drawing.

5.2 Using Copy, Move, Mirror, and Rotate

The following tutorial includes an example of the use of copy, move, mirror, and rotate commands to create an architectural layout for two bathrooms.

1. Open AutoCAD 2024 from the Desktop shortcut.
2. Choose Open from the Quick Access toolbar.
3. Choose the **Bath Layout.dwg** drawing file from the **AutoCAD 2024 Certified User Exercises Files \ Exercise Files \ Ch 5** folder.
4. Choose the **Move** command and move the table shown at p1 to the end of the line shown at p2 in Figure 5-46.

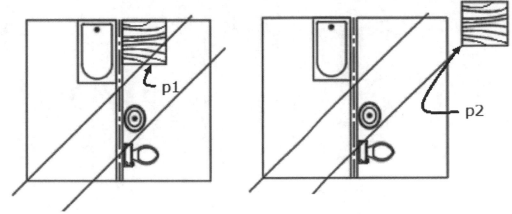

Figure 5-46

5. Choose the **Copy** command and copy the bathtub shown at p1 to the corner as shown at p2 in Figure 5-47.

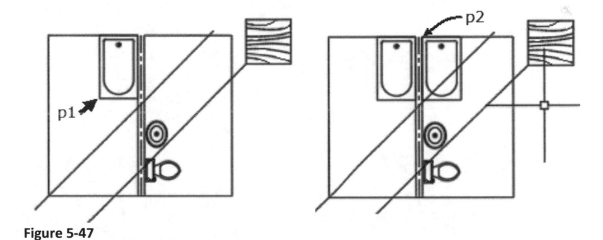

Figure 5-47

6. Choose the Mirror command and mirror copy the plumbing fixtures shown at p1 and p2 as shown in Figure 5-48. You may use the center line shown at p3 and p4 as the mirror axis line. Upon completion of mirror operation, the fixtures should be placed as shown at p5 in Figure 5-48.

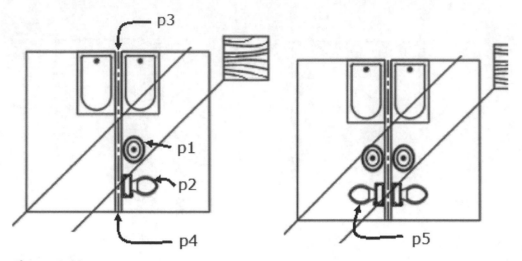

Figure 5-48

7. Choose the Rotate command to rotate the bathroom layout fixtures and line about the rotation basepoint, as shown at p1 in Figure 5-49. Rotate the content to match the angle of the line shown at p2. The completed bath layout is shown at p3 in Figure 5-49.

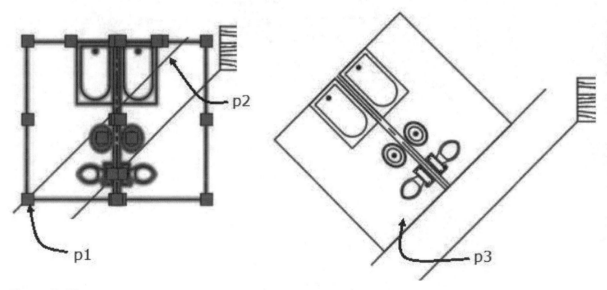

Figure 5-49

8. Save the drawing as **Bath_Layouts_Your_Name** to your student folder and close the drawing.

5.3 Using the Enhanced Fillet – Chamfer – Blend Commands

The following tutorial includes the use of the enhanced fillet and chamfer tools.

1. Open AutoCAD 2024 from the Desktop shortcut.
2. Choose Open from the Quick Access toolbar.
3. Choose the **Fillet Chamfer.dwg** drawing file from the **AutoCAD 2024 Certified User Exercises Files \ Exercise Files \ Ch 5** folder.
4. Select the **Fillet** command from the Modify panel, as shown in Figure 5-50.

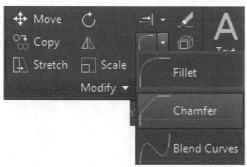

Figure 5-50

5. Respond to the workspace prompts as shown below.

   Select first object or: (Select the line at p1 as shown in Figure 5-51.)

   Select second object or shift-select to apply corner or [Radius]: r (Press the down arrow key to display the command options. Continue to press the down arrow key to toggle to the **Radius** option; press **Enter**.)

   Specify fillet radius <0.0000>: .5 (Type **.5**, press **ENTER**.)

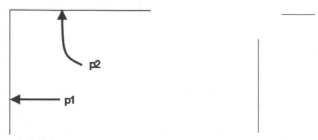

Figure 5-51

Select second object or shift-select to apply corner or [Radius]: (Move the cursor to the line at p2 to display a preview of the radius as shown in Figure 5-52. After viewing the preview select the line at p2.)

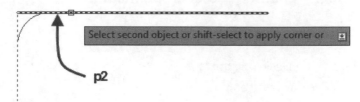

Figure 5-52

6. The completed fillet is shown in Figure 5-53.

Figure 5-53

7. In this step you will create a corner intersection of two lines. Choose the Fillet command, respond to the workspace prompts as shown below.

Command: _fillet

Current settings: Mode = TRIM, Radius = 0.5000

Select first object or [Undo/Polyline/Radius/Trim/Multiple]: (Select the line at p1 as shown in Figure 5-54.)

Select second object or shift-select to apply corner or [Radius]: (Press **Shift**, select the line at p2 in Figure 5-54.)

The corner intersection is shown at right in Figure 5-54.

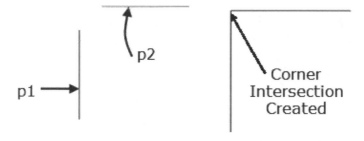

Figure 5-54

8. Choose the Chamfer command of the Modify panel.

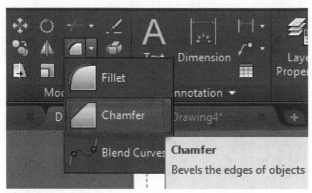

Figure 5-55

9. Respond to the workspace prompts as shown below.

   Command: _chamfer

   (TRIM mode) Current chamfer Dist1 = 0.0000, Dist2 = 0.0000

   Select first line or: **d** (Press the down arrow key to display the command options.
   Continue to press the down arrow key to toggle to the **Distance** option, press **Enter**.)

   Specify first chamfer distance <0.0000>: **.25** (Type **.25** to set the distance.)

   Specify second chamfer distance <0.2500>: **.5** (Type **.5** to set the second distance.)

   Select first line or [Undo/Polyline/Distance/Angle/Trim/mEthod/Multiple]: (Select the
   line at p1 as shown in Figure 5-56.)

   Select second line or shift-select to apply corner or [Distance/Angle/Method]: (Select
   the line at p2 as shown in Figure 5-56.)

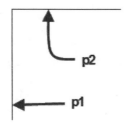

Figure 5-56

The chamfer formed is shown in Figure 5-57.

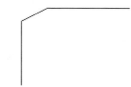

Figure 5-57

10. Choose the Blend command from the Modify panel as shown in Figure 5-58.

Figure 5-58

11. In this step, you will create a blend between the two horizontal lines as shown at right in the Fillet Chamfer. Respond to the workspace prompts as shown below.

    Command: _BLEND

    Continuity = Tangent

    Select first object or [CONtinuity]: (Select the line at **p1** as shown in Figure 5-59.)

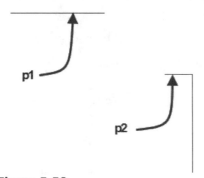

Figure 5-59

Select second object: (Move the cursor over the line at **p2** to preview the blend as shown in Figure 5-60. After viewing the preview select the line at **p2.)**

The completed Blend is shown at right in Figure 5-60.

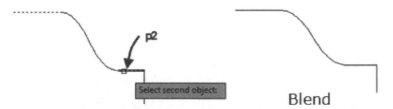

Figure 5-60

12. To invert the blend select the blend to display its grips as shown in Figure 5-61.

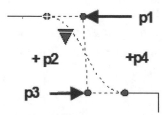

Figure 5-61

Select the grip at p1 and drag the grip to the location shown at **p2** in Figure 5-61.

Select the grip at **p3** and drag the grip to **p4** as shown in Figure 5-61.

Press Escape to clear the grip selection.

The inverted blend is shown in Figure 5-62.

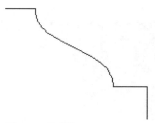

Figure 5-62

13. Save the drawing as **Fillet_Chamfer _Blend_Your Name** in your student folder and close the drawing.

5.4 Using Offset, Trim, and Extend

The following tutorial includes exercises in the use of the offset command to both draw and measure distances to create lines. After creating the lines, you will experience editing with trim and extend commands.

1. Open AutoCAD from the Desktop shortcut.
2. Choose **QNew** from the Quick Access toolbar.
3. Choose **acad.dwt** from the Select Template dialog box as shown in Figure 5-63. Choose **Open** to select the template.

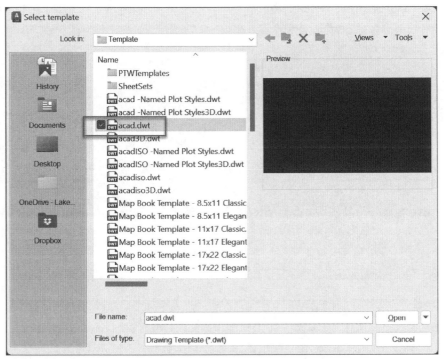

Figure 5-63

4. Verify **Drafting & Annotation** is the current workspace as shown in the Quick Access toolbar as shown in Figure 5-64. If necessary, choose Drafting & Annotation from the workspace flyout of the Quick Access toolbar.

Figure 5-64

5. Choose the Application Menu (red A in the upper left corner of the workspace). Choose the flyout at p1 to select Drawing Utilities > Units from the Application Menu, as shown in Figure 5-65.

Figure 5-65

6. The Drawing Units dialog box opens; edit the Length Type to Architectural as shown in Figure 5-66. Choose OK to dismiss the dialog box.

Figure 5-66

7. Choose the Customization toggle of the Status bar to display the flyout toggle list. Choose the Units toggle as shown in Figure 5-67. The Units of the drawing are displayed in the Status bar as shown below. You may change the units by selecting from the flyout of the Units toggle.

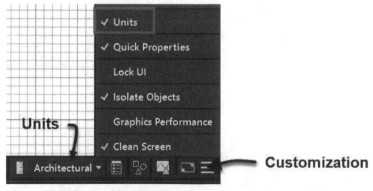

Figure 5-67

8. Click the flyout of the Grid toggle of the Status bar; choose Snap Settings as shown at p1 in Figure 5-68.

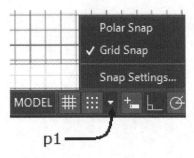

Figure 5-68

9. Edit the Drafting Settings dialog box as shown in Figure 5-69. Do not press Enter, simply click in a field and continue to click or press **Tab** to edit the fields. (Pressing Enter will dismiss the dialog box.) Click the Snap spacing, type **8**. Click in the Snap Y spacing and verify **8** is displayed. Click the Grid X spacing type **8**. Click the Grid Y spacing, verify **8** is set. Edit the Major line every to **6**. Verify Snap ON (F9) and Grid (F7) are toggled ON. Choose **OK** to dismiss the dialog box.

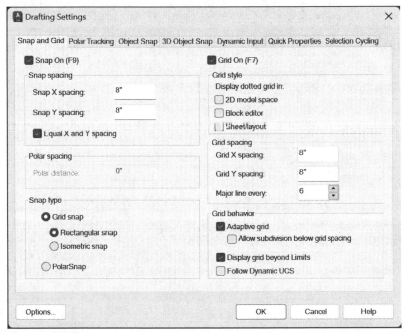

Figure 5-69

10. Click in the command window and type **LIMITS,** press ENTER, respond to the command prompts as shown below.
    Command: LIMITS
    Specify lower left corner or [ON/OFF] <0'-0",0'-0">: **(Press ENTER)**
    Specify upper right corner <1'-0",0'-9">: Type **30',30',** press **ENTER**

11. Choose Zoom All from the Zoom flyout of the Navigation bar at right as shown in Figure 5-70.

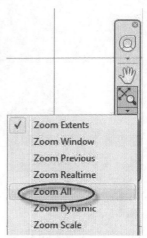

Figure 5-70

12. When you choose the Zoom All command, the screen will display an area of approximately 30'x30'. The results of the grid changes provide a grid line every 8".

13. Move the cursor to the workspace. Notice the cursor includes a square pick box since no command is active, as shown in Figure 5-71.

Cursor with pickbox

Figure 5-71

14. Choose the Home tab. Choose the **Line** command of the Draw panel.

15. Move the cursor to the workspace. Notice the square pick box has been replaced by draw mode cursor as shown in Figure 5-72 since you have activated a draw command.

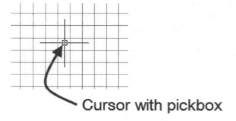

Specify first point:

Figure 5-72

16. The Snap and Grid setting will lock the cursor movement to 8" multiples. In this tutorial you will create a patio which will be covered with brick pavers that are 4" x 8"; the snap control will reduce cutting of pavers.

17. Edit the toggles of the Status bar as shown in Figure 5-73.

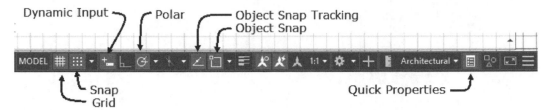

Figure 5-73

18. The **Polar** toggle will lock the cursor movement angle. The **Object Snap** toggle will lock the cursor to specific locations on lines or other entities and the **Dynamic Input** toggle will display the prompts for the command in the workspace as shown in Figure 5-74.

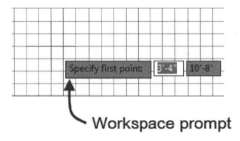

Figure 5-74

19. The display of the coordinates of the cursor can be displayed in the Status bar. Choose Customization of the Status bar, shown in Figure 5-75, to display the toggle list. Choose the **Coordinate** toggle from the top of the list.

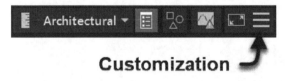

Figure 5-75

20. Move the cursor to the lower left corner of the screen. Within this region is the coordinate display as shown in Figure 5-76. Move the cursor to the location **0, 4'**; left click to begin the line.

Figure 5-76

21. Move the cursor to the right to display a polar distance and angle of **30'<0** as shown in Figure 5-77; left click to specify the endpoint.

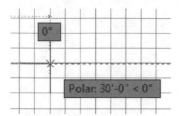

Figure 5-77

22. Continue in the Line command by moving the cursor up to display a polar angle of **90** degrees as shown in Figure 5-78. Notice the distance that you have moved up is displayed in blue highlight; type **16'** to edit this dynamic dimension.

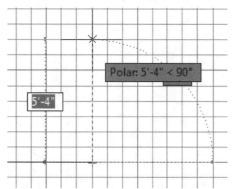

Figure 5-78

23. Toggle OFF the **Grid** toggle in the Status bar to remove the display of the grid. Toggle OFF the Snap toggle to remove the cursor control.

24. The two lines created should be displayed in the workspace, as shown in Figure 5-79.

Figure 5-79

25. In the next steps, we will complete the rectangle by using the Offset command. In this sequence you will use *Noun Verb* selection since you will select the line then select the **Offset** command. Select the 30' line and choose the **Offset** from the Modify panel in the Home tab, as shown in Figure 5-80.

Figure 5-80

26. Respond to the workspace prompts as shown below:

    Command: _offset

    Specify offset distance or [Through/Erase/Layer] <Through>: **16'**

    Specify point on side to offset or [Exit/Multiple/Undo] <Exit>: (Select a point above the line as shown at p1 in Figure 5-81.)

    Specify point on side to offset or [Exit/Multiple/Undo] <Exit>: (Press **ESC** key to end the command.)

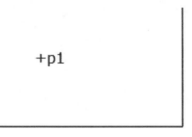

+p1

Figure 5-81

27. In this sequence you will use *Verb Noun* selection since you will choose the Offset command then select the line for the offset. Press the **Space** bar to repeat the last command.

28. Respond to the workspace prompts as shown below:

    Command: OFFSET

    Specify offset distance or [Through/Erase/Layer] <16'-0">: **30'** Press **ENTER**

    Select object to offset or [Exit/Undo] <Exit>: (Select the line as shown at p1 in Figure 5-82.)

    Specify point on side to offset or [Exit/Multiple/Undo] <Exit>: (Move the cursor to the left and click as shown at p2 in Figure 5-82.)

    Select object to offset or [Exit/Undo] <Exit>: (Press **ESC** to end the command.)

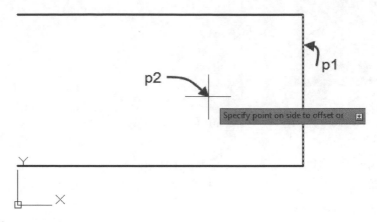

Figure 5-82

29. Next you will create 4'-0" planters at the left and right of the patio as shown in Figure 5-83.

Figure 5-83

30. Choose the Offset command from the Modify panel in the Home Tab.

31. Respond to the workspace prompts as shown below.

Command: _offset

Current settings: Erase source = No Layer=Source OFFSETGAPTYPE=0

Specify offset distance or [Through/Erase/Layer] <30'-0">: **4'** (Type **4'** press **Enter** to specify the offset distance.)

Select object to offset or [Exit/Undo] <Exit>: (Select the line at p1 shown in Figure 5-84.)

Select object to offset or [Exit/Undo] <Exit>: (Select a point near p2 as shown in Figure 5-84.)

Specify point on side to offset or [Exit/Multiple/Undo] <Exit>: (Press Escape to end the command.)

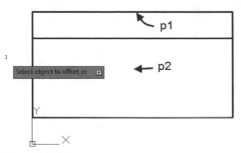

Figure 5-84

32. Right-click and choose **Repeat OFFSET** from the short-cut menu as shown in Figure 5-85.

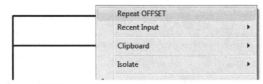

Figure 5-85

33. Respond to the workspace prompts as shown below.

Command:

OFFSET

Specify offset distance or [Through/Erase/Layer] <4'-0">: (Press **ENTER** to accept the default value.)

Select object to offset or [Exit/Undo] <Exit>: (Select the line at p1 as shown in Figure 5-86.)

Specify point on side to offset or [Exit/Multiple/Undo] <Exit>: (Move the mouse to a point near p2 and left click to copy the line as shown in Figure 5-86.)

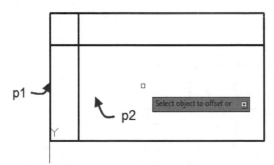

Figure 5-86

Continue from the previous sequence, respond to the following workspace prompts:

Select object to offset or [Exit/Undo] <Exit>: (Select the line at p3 as shown in Figure 5-87.)

Specify point on side to offset or [Exit/Multiple/Undo] <Exit>: (Select a point near p4 as shown in Figure 5-87.)

Select object to offset or [Exit/Undo] <Exit>: (Press Escape to end the command.)*Cancel*

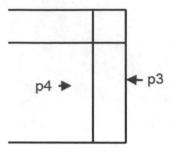

Figure 5-87

34. In this step you will remove a portion of the lines using the trim command. Select **Trim** from the Modify panel as shown in Figure 5-88.

Figure 5-88

35. Respond to the workspace prompts as follows:

Command: _trim

Select object to trim or shift-select to extend or: (Select the line at p1 as shown in Figure 5-89.)

Select object to trim or shift-select to extend or: (Press ESC to end the trim command.)

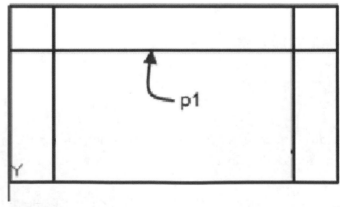

Figure 5-89

36. The patio now includes the following lines as shown in Figure 5-90.

Figure 5-90

37. Refer to previous command sequences and repeat the Trim command to remove the lower portion of two lines as shown in Figure 5-91.

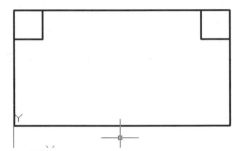

Figure 5-91

38. Refer to previous command sequences to use the Offset command to create lines to represent a retaining wall 1' wide at the right and top of the patio as shown in Figure 5-92.

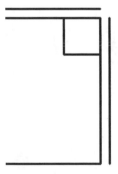

Figure 5-92

39. In this step you will use the Fillet command to extend the two exterior wall lines to intersection. Select the Fillet command from the Modify panel in the Home tab. Respond to the command prompts as listed in the Command Window as follows:
Command: _fillet
Current settings: Mode = TRIM, Radius = 0'-0"

Select first object or [Undo/Polyline/Radius/Trim/Multiple]: (Select the line at p1 as shown in Figure 5-93.)

Select second object or shift-select to apply corner or [Radius]: (Hold the Shift key down and select the line at p2 as shown in Figure 5-93.)

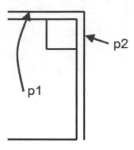

Figure 5-93

Repeat the Fillet command add lines that end the retaining walls (as shown in Figure 5-94).

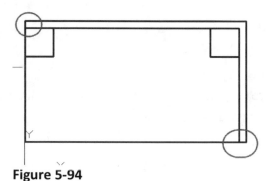

Figure 5-94

40. Choose Layout 1 in the lower left corner as shown in Figure 5-95.

Figure 5-95

41. Double click inside the viewport at p1 as shown in Figure 5-96.

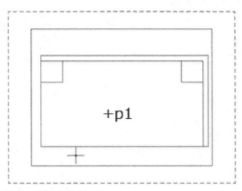

Figure 5-96

42. Select the Viewport scale flyout to choose the 1/8" = 1'-0" as shown in Figure 5-97.

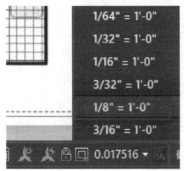

Figure 5-97

43. The patio should appear in the viewport as shown in Figure 5-98.

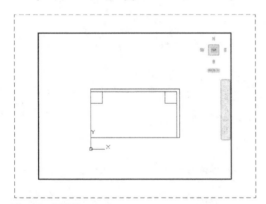

Figure 5-98

44. Choose **Save** from the **Quick Access toolbar**. Since the drawing has not been named, type **Offset Trim Extend_Your Name** as the drawing name and save the drawing to your personal folder.

5.5 Using the Stretch Command

The following tutorial includes use of the Stretch command.

1. Verify **Drafting** & **Annotation** is the current workspace as shown in the Quick Access toolbar.
2. Choose Open from the Quick Access toolbar.
3. Navigate to the AutoCAD 2024 Certified User Exercise \ Exercise Files \ Ch 5 folder. Select the **Stretch**.dwg.
4. Verify **Dynamic Input** is toggled on in the Status bar. Toggle off Object Snap and toggle ON Polar.
5. Choose **Stretch** from the Modify panel of the Home tab.

   Respond to the workspace prompts as shown below.

   Command: _stretch

   Select objects to stretch by crossing-window or crossing-polygon...

   Select objects: *(Select a point at p1 as shown* in Figure 5-99.)

   Specify opposite corner: *(Select a point at p2 as shown* in Figure 5-99.)

   2 found

   Select objects: *(Press Enter to end selection.)*

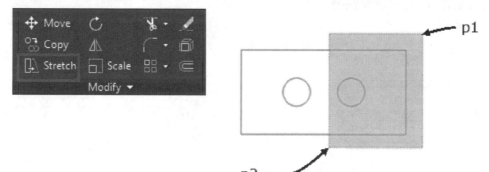

Figure 5-99

Specify base point or [Displacement] <Displacement>: (Select a point at **p1** as shown in Figure 5-100.)

Specify second point or <use first point as displacement>: 1 (Move the cursor to **p2** to display a polar angle of 0; type **1** to stretch the rectangle 1" as shown in Figure 5-100.)

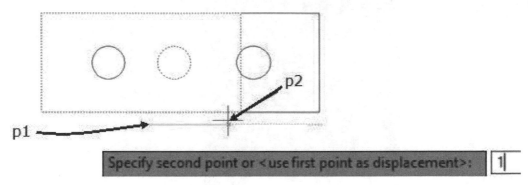

Figure 5-100

The entities are stretched as shown in Figure 5-101.

Figure 5-101

6. Save the drawing to your student folder. Name the file to **Stretch _ Your Name**.
   Question:
   Which of the following selection methods may be used when selecting entities using the
   Stretch command?
   1. Fence
   2. Window
   3. Window Polygon
   4. Crossing Window
   5. Crossing Polygon
   6. Last
   7. All

5.6 Join and Break Commands

1. Open AutoCAD 2024 from the Desktop shortcut.
2. Choose Open s from the Create tab.
3. Navigate to the AutoCAD 2024 Certified User Exercise Files \ Exercise Files \Ch 5 folder.
   Choose the **Join Break.dwg**.
4. Verify Dynamic Input is toggled on in the Status bar.
5. Toggle Off the Grid. Click the Object Snap toggle of the Status bar and verify the Nearest
   and Intersection modes are toggled ON as shown in Figure 5-102.

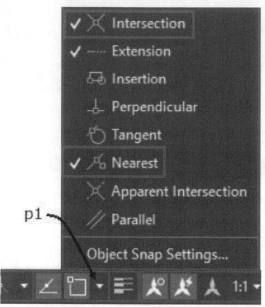

Figure 5-102

6. In the next series of steps, you will join two line segments as shown at p1 and p2 in Figure 5-103.

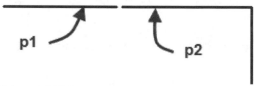

Figure 5-103

7. Choose the Home tab. Expand the Modify panel, choose Join as shown in Figure 5-104.

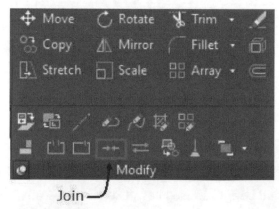

Figure 5-104

8. Respond to the Workspace prompts as follows:
Command: _join

Select source object or multiple objects to join at once: (Select the line at p1 shown in Figure 5-105.) 1 found
Select objects to join: (Select the line at p2 as shown in Figure 5-105.) 1 found, 2 total
Select objects to join: **Enter** (Press **Enter** to end selection.)
2 lines joined into 1 line

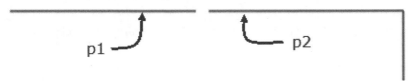

Figure 5-105

9. In the next series of steps, you will use the Break command to remove a portion of a line to clearly display text and edit the arc using the trim command

10. Extend the Modify panel, choose the Break tool as shown in Figure 5-106.

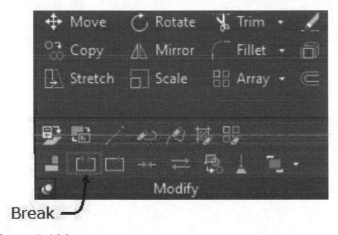

Figure 5-106

11. Respond to the workspace prompts as follows:
    Command: _break
    Select object: (Select the line at p1 as shown in Figure 5-107.)
    Specify second break point or [First point]: (Select the line at p2 as shown in Figure 5-107.)

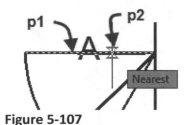

Figure 5-107

12. Choose the Trim command from the Modify panel.

    Respond to the workspace prompts as follows:

    Command: _Trim

    Current settings: Projection=UCS, Edge=None, Mode=Quick

    Select cutting edges ...1 found

    Select object to trim or shift-select to extend or: (Select the arc at p1 as shown in Figure 5-108.).

    Select object to trim or shift-select to extend or: (Press **Escape** to end the command.)

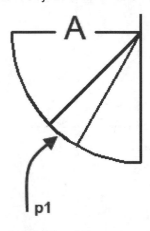

Select object to trim or shift-select to extend or

Figure 5-108

13. The arc segment is removed from p1 to p2 as shown in Figure 5-109.

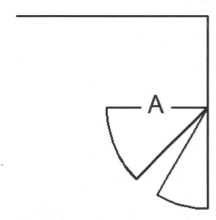

Figure 5-109

14. Save the drawing as **Join_Break _ Your name** in your student folder and close the drawing.

5.7 Manipulating Entities with Grips

The following tutorial includes the use of grips to copy, rotate, and mirror entities.

1. Launch AutoCAD 2024 from the desktop. Start a new drawing using the acad.dwt template.
2. Verify that the Drafting & Annotation workspace is current in the Quick Access toolbar.
3. Verify Polar, Object Snap, and Dynamic Input are toggled On in the Status bar.
4. Choose **Line** from the Draw panel of the Home tab. Respond to the workspace prompts as follows.
    LINE
    Specify first point: **2,2** (Type 2,**2** and press enter in the workspace prompt as shown in Figure 5-110.)

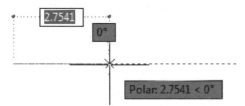

Figure 5-110

Move the cursor to the right to display the dynamic dimension as shown in Figure 5-111.

Figure 5-111

Type **2** to enter the length. Press **Tab** to enter the angle as shown in Figure 5-112.

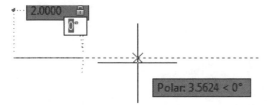

Figure 5-112

Type **0** then **ENTER** to specify the polar angle.
Specify next point or [Undo]: (Press **ENTER** to end the command.)

5. In this step you will copy the 2" line to a specified angle using Rotate Copy. Select the line to display its grips as shown in Figure 5-113.

Figure 5-113

Select the grip at p1 as shown Figure 5-114 below. Right-click choose **Rotate** from the contextual menu shown in Figure 5-114.

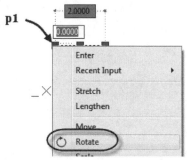

Figure 5-114

Retain the cursor location at p1, right-click and choose **Copy** from the contextual menu shown in Figure 5-115.

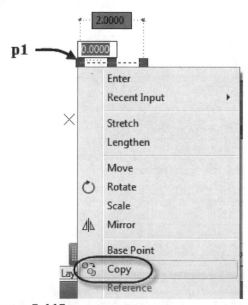

Figure 5-115

Press **Tab** to specify the angle of rotation as shown in Figure 5-116. Type **120** in the dynamic dimension.

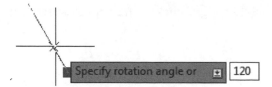

Figure 5-116

Press Escape to clear the grips; the line is copied to its new position as shown in Figure 5-117.

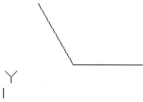

Figure 5-117

6. Repeat the procedure of step 5 to rotate the horizontal 2" line -120° as shown in Figure 5-118.

Figure 5-118

7. In the next series of steps, you will mirror the three lines about a horizontal axis. Select the three lines using a crossing selection; left click at p1 then p2 as shown in Figure 5-119.

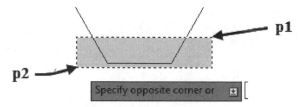

Figure 5-119

8. Hold the shift key down, select the grips at p1, p2, and p3 as shown in Figure 5-120.

Figure 5-120

9. Left-click the grip shown at p1 in Figure 5-121, right-click then choose **Mirror** from the context menu.

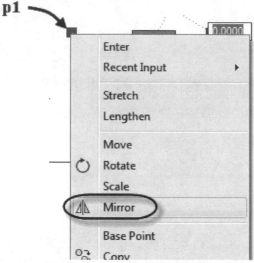

Figure 5-121

10. Retain the cursor at p1, right-click and choose **Copy** from the context menu to copy the lines during the mirror operation. Respond to the following workspace prompts:
** MIRROR (multiple) **
Specify second point or [Base point/Copy/Undo/eXit]: (Select the grip at p2 as shown in Figure 5-122.)

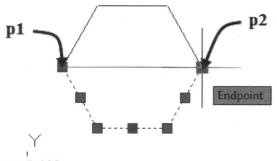

Figure 5-122

11. Press **Escape** twice to end the edit and clear grip selection.
12. In the next series of steps, you will copy stretch the lines of the polygon 5" from the basepoint and use the CTRL key to apply the snap distance to additional copy stretch operations.
13. Select the lines of the polygon to display the selection grips. Hold the **Shift** key down, choose the grips at p1, p2, p3, p4, p5 and p6 as shown in Figure 5-123.

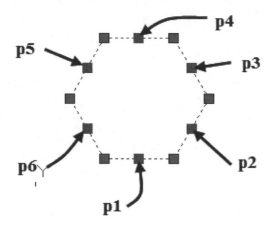

Figure 5-123

14. Left click the grip at p1 to make it hot and right-click and choose **Copy** from the context menu. Move the cursor up; type **5** in the dynamic dimension as shown in Figure 5-124.

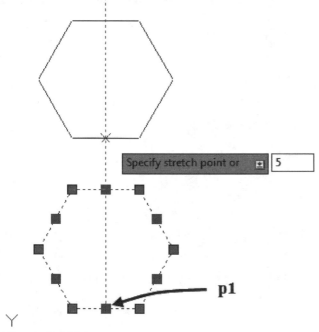

Specify stretch point or ▣ 5

p1

Figure 5-124

15. Hold the **Ctrl** key down to retain the grip snap value, move the cursor to the right. Click near p1 as shown in Figure 5-125 to copy an instance of the lines to the right.

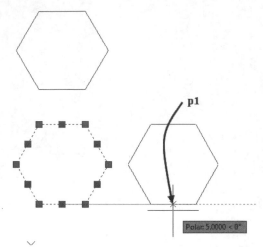

Figure 5-125

16. Continue to hold the **Ctrl** key down, move the cursor to p1 as shown in Figure 5-126 and click to copy the polygon 5 units up and right using the grip snap value.

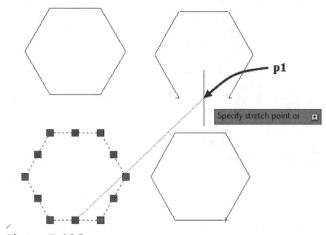

Figure 5-126

17. Press Escape twice to end the grip edit and clear the selection of grips.
18. Save the file as **Grips Your Name** in your student folder and close the file.

5.8 Grips Quiz

Name_____

1. When you select an entity in AutoCAD the object highlights and its grips are displayed in default color of _____.
2. When you have selected an entity and move your cursor to hover over the displayed grip the grip changes to the color _____.
3. When you select a grip its color changes by default to the color of _____.
4. In order to select the two grips shown below you must first press and hold the _____ key prior to selecting the first grip and the second grip.
5. If you select a line to display its grips then select the endpoint grip you can stretch the line; however, if you want to rotate the line you must either right-click and Rotate or press the _____ of the keyboard to cycle to Rotate.
6. To end grip editing press the _____ key of the keyboard.
7. To remove the display of grips for an entity press the _____ key of the keyboard.
8. To activate a grip offset snap value when copying a line hold down the _____ key after creating the first copy.
9. The line shown in Figure 5-127 is 15" long. To increase the length to 16" without changing the angle, choose _____ from the dynamic menu shown below. To toggle between Stretch and Lengthen press the _____ key of the keyboard.

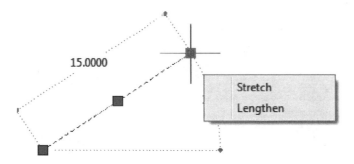

Figure 5-127

Notes:

Chapter 6
Using Additional Drawing Techniques

This chapter will review the methods of creating and editing arrays of entities using Copy, Rectangular Array, Path Array, and Polar Array commands. The chapter also includes a review of the Hatch command and editing of existing hatches. The island detection methods and gap tolerance are included in the Option settings when placing a hatch. The applications and benefits of annotative and associative hatches are included in tutorials. Finally, methods to create and edit polylines are presented. There are six tutorials at the end of the chapter.

The student will be able to:

1. Create an array using the Array option of the Copy command
2. Create a rectangular array using Array Creation contextual tab and grips
3. Create a path array using the distance and measure options
4. Create multiple levels of arrays
5. Create associative and non-associative arrays
6. Create hatch patterns using various patterns, scales with annotative and associative properties
7. Create and edit polylines using the width, curve fit, join and spline options and the Pedit command

Copy Array

The Copy command includes an array option that allows you to create a one-dimensional array by entering the number of items and the distance between items or the number of items to fit between two points. The technique to create an array of entities using the Copy command with array option is shown below. Note you can open the **Copy Array demo.dwg** drawing file from the **AutoCAD 2024 Certified User Exercise Files \ Exercise Files \ Certification Demo Files \ Ch 6** folder to try this technique as you read.

Toggle ON Geometric Center object snap in the Status bar.

Choose **Copy** from the Modify panel.

Select the polygon located in the Copy Array demo drawing.

Specify base point or [Displacement/mOde] <Displacement>: _gcen of (Move cursor over the polygon; when the Geometric Center marker displays click the marker as shown at p1 in Figure 6-1.)

Specify second point or [Array] <use first point as displacement>: **A** (Press the **down arrow** to display the command options menu, continue to press the key to toggle to **Array** and press **Enter** to select as shown at p2 in Figure 6-1.)

Enter number of items to array: **5** (Type **5** in the dynamic dimension field shown at p3 in Figure 6-1.)

Specify second point or [Fit]: **2'** (Type **2'** press **ENTER** to specify the distance between entities of the array as shown at p4 in Figure 6-1.)

Specify second point or [Array/Exit/Undo] <Exit>: (Press **ENTER** to exit the command. The complete array shown at p5 in Figure 6-1.)

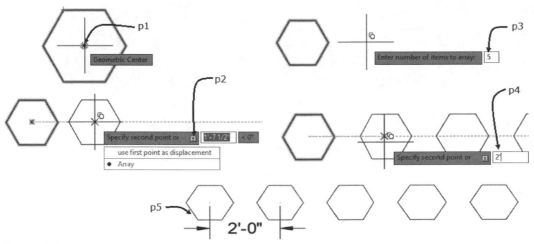

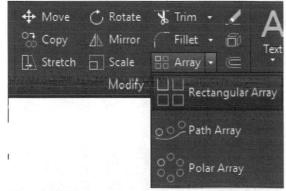

Figure 6-1

Rectangular Array

The Array command is located on the Modify panel in the Home tab. The Array command flyout includes Rectangular Array, Path Array, and Polar Array options.

Figure 6-2

Note: you can open the **Array demo.dwg** drawing file from **AutoCAD 2024 Certified User Exercise Files \ Exercise Files \ Certification Demo Files \ Ch 6** folder to try the rectangular array command as follows. When you choose the Rectangular Array command and select an entity, the Array Creation tab opens, as shown in Figure 6-3. The Array Creation tab includes panels for setting the row and column distances and number of items. The Rectangular Array can be specified by editing the Array Creation Tab or the grips. The insertion point of the array is shown at p1 in Figure 6-3. Note that a positive column distance develops the array to the right, and a negative column distance develops the array to the left of the insertion point. A positive row distance develops the array up from the insertion point, as shown in the current settings, and a negative row distance develops the array down from the insertion point. The grips of the array allow the distances to be changed between rows as shown at p2 in Figure 6-3.

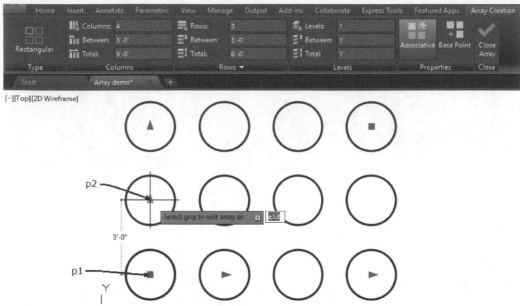

Figure 6-3

The Properties panel includes the Associative button as shown in Figure 6-3. Associative arrays associate all the settings in one object; therefore, when you choose the Close Array button the tab closes and the array settings are retained. If you select an entity of an existing associative array, all settings for the array edit are displayed in an Array tab shown in Figure 6-4 or in Quick Properties. If you select an entity of a non-associative array, no Array tab opens—you will only edit the selected entity.

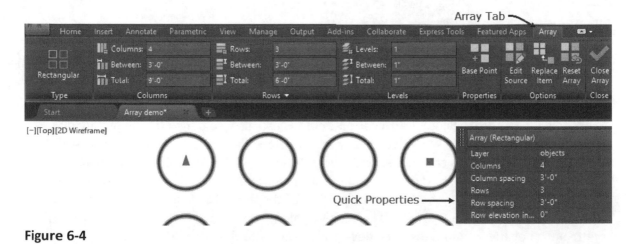

Figure 6-4

Path Array

The Path Array command allows you to create copies of one or more entities distributed along a path. The path can be a line, polyline, spline, arc, or circle. The entities are evenly distributed along the path using the Divide option or distributed at a specific distance along the path using the Measure option. The entities can be aligned with the path or specified in the Z Direction.

The technique to create a path array is shown below. Note you can open the **Path Array demo.dwg** drawing file from the **AutoCAD 2024 Certified User Exercise Files \ Exercise Files \ Certification Demo Files \ Ch 6** folder to try this technique as you read the steps.

Select the polygon as shown at p1 in Figure 6-5.
Choose Path Array command from the Modify panel in the Home tab.
Select path curve: (Select the arc as shown at p2 in Figure 6-5.)
Select grip to edit array or: (The Array Creation contextual tab opens, verify the Associative, Align Items and Z Direction are toggled on in the Properties panel. Type 2' 6" in the Between field of the Items panel in Array Creation tab as shown at p3 in Figure 6-5.)
Select grip to edit array or: T (Press the down arrow key to display the Options menu, toggle to the Tangent option, press Enter as shown at p4 in Figure 6-5.)
Specify first point of tangent direction vector or [Normal]: (Using the Nearest object snap mode choose a point on the arc as shown at p5 in Figure 6-5.)
Specify second point of tangent direction vector: (Using the Nearest object snap mode choose a point on the arc as shown at p6 in Figure 6-5.)
Select grip to edit array or: (Choose Close Array to dismiss the Array Creation tab and view the completed path array in Figure 6-6.)

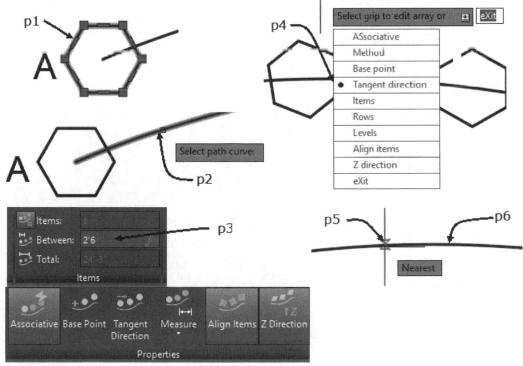

Figure 6-5

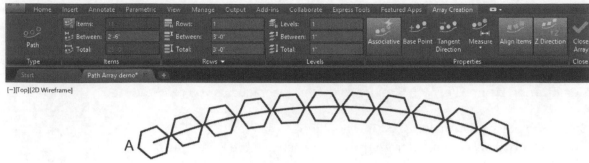

Figure 6-6

Polar Arrays

Polar arrays are used to copy an entity in a circular pattern about a center point. The Array Creation tab shown in Figure 6-7 includes an Items panel that allows you to specify the number of items or the angle between items with fill angle. The Properties panel allows you to control if entities are rotated. The default rotation of an entity is counterclockwise from the basepoint. The Direction button allows you to specify clockwise or counter-clockwise rotation from the basepoint.

The technique to create a polar array 135 degrees from the basepoint is described below. Note you can open the **Polar Array demo** file from the **AutoCAD 2024 Certified User Exercise Files \ Exercise Files \ Certification Demo Files \ Ch 6** folder to try this technique as you read the steps.

Select the polygon shown at p1 in Figure 6-7.
Choose the Polar Array command from the Modify panel as shown at p2 in Figure 6-7.
Command: _arraypolar 1 found
Type = Polar Associative = Yes
Specify center point of array or: (Using the intersection object snap mode select the intersection as shown at p3 in Figure 6-7.)
Select grip to edit array or: (In the Array Creation tab, edit the Fill field to **135** as shown at p4 and the Items to 3 at p5 in the Items panel shown in Figure 6-7.)
Select grip to edit array or: (In the Array Creation tab, toggle the Direction off as shown at p6 in Figure 6-7 to set the rotation to **clockwise**.)
Select grip to edit array or: (Choose the **Close Array** button as shown at p7 in Figure 6-7 to end the command.)

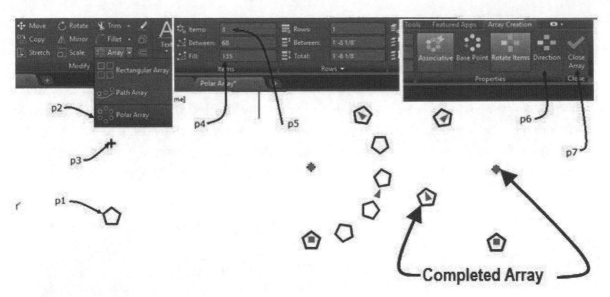

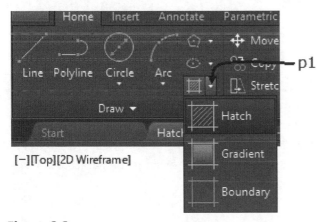

Figure 6-7

Hatching Overview

Hatching is added to a pattern to represent materials cut in sections and in elevations. The Hatch command flyout shown in Figure 6-8 includes Hatch, Gradient, and Boundary commands.

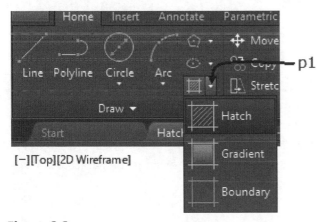

Figure 6-8

When you choose the Hatch command, the Hatch Creation contextual tab opens as shown in Figure 6-9.

Figure 6-9

The Boundaries panel includes the Pick Points and Select buttons. The Pick Points button prompts you to select an internal point. From this point the software floods the area and determines the boundary as shown at p1 in Figure 6-10, whereas the Select button prompts you to select objects such as circles, polygons or rectangles as shown at p2 in Figure 6-10. The Options panel shown in Figure 6-10 includes an Annotative button which if turned On the scale of the hatch is controlled by the annotation scale in the Model tab or by the Viewport Scale in a layout. The Options panel extension as shown at p3 allows you to select how the software treats islands when multiple objects are selected. The Normal Island Detection was used to create the hatch shown at p2 in Figure 6-10.

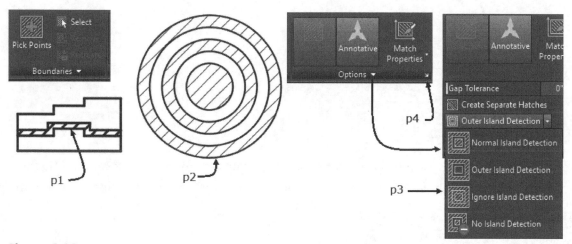

Figure 6-10

Click the arrow shown at p4 in Figure 6-10 to open the Hatch and Gradient dialog box shown in Figure 6-11 to edit island detection and additional settings for controlling hatch placement. Note the graphic examples of island detection are shown at p1 in Figure 6-11. You can set the **gap tolerance** as shown at p2 in Figure 6-11 to specify a gap tolerance distance, which allows a hatch to be created from boundaries with gaps; if the gap does not exceed the gap tolerance distance specified the hatch will be applied.

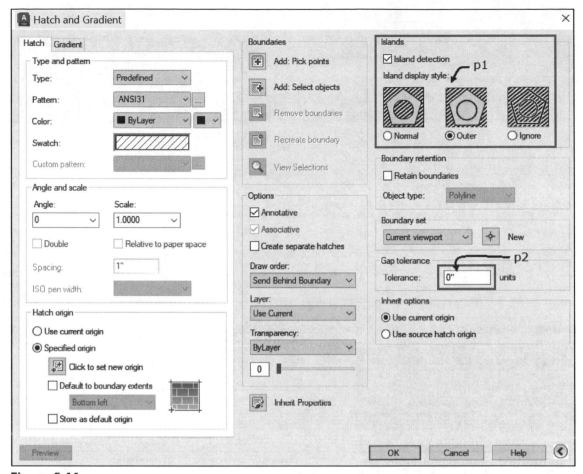

Figure 6-11

The technique to use the Hatch command is shown in the following workspace prompts and Figure 6-12. Note that you can open the **Hatch demo.dwg** drawing file from the **AutoCAD 2024 Certified User Exercise Files \ Exercise Files \ Certification Demo Files \ Ch 6** folder to try this technique as you read the steps.

Choose Hatch from the Draw panel to open the Hatch Creation tab.
Choose ANSI31 pattern from the Pattern panel as shown at p1 in Figure 6-12.
Verify **Annotative** is turned ON in the Properties panel as shown at p2 in Figure 6-12.
Choose the **Pick Points** button of the Boundaries panel as shown at p3 in Figure 6-12.

Pick internal point or: Move the cursor to near point p4, click to place the hatch and choose **Close Hatch Creation** tab as shown at p5 to complete the command.

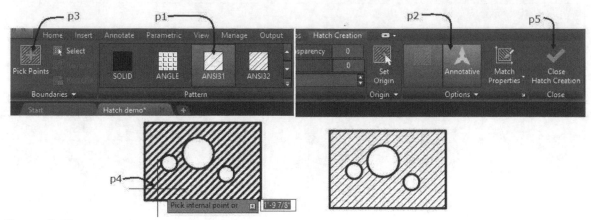

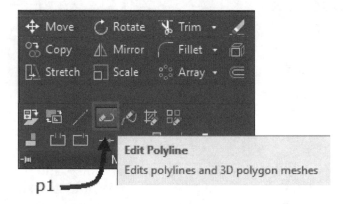

Figure 6-12

Editing Polylines

Polylines can be edited using the PEDIT command and Properties palette. Choose PEDIT command from extended Modify panel as shown at p1 in Figure 6-13.

Figure 6-13

Converting Lines to a Polyline and a Curve Fit

The technique to use the Pedit command is shown in the following workspace prompts and Figure 6-14. Note that you can open the **Pedit demo** file from the **AutoCAD 2024 Certified User Exercise Files \ Exercise Files \ Certification Demo Files \ Ch 6** folder to try this technique as you read the steps.

Choose the Pedit command from the Modify panel shown at p1 in Figure 6-13.

Command: PEDIT

Select polyline or [Multiple]: (Select the line shown at **p1** in Figure 6-14.)

Object selected is not a polyline

Do you want to turn it into one? <Y> (Press **Enter** to accept the Yes default)

Enter an option: J (Press the down arrow key to toggle to the **Join** option as shown at p3 in Figure 6-14; press **Enter** to choose Join.)

Select objects: Specify opposite corner: 3 found (Select a point at p4 to p5 as shown in Figure 6-14.)
Select objects: (Press **Enter** to end the command.)
3 segments added to polyline

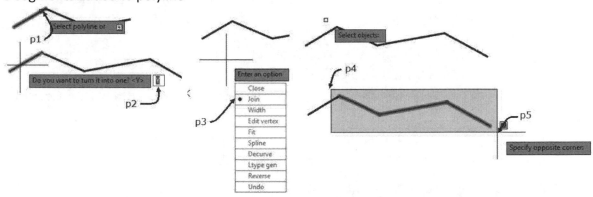

Figure 6-14

Select the polyline, right click and choose Polyline > Curve Fit as shown at p1 of the contextual menu in Figure 6-15. The resulting curve which has points common to the vertex of the polyline is shown at p2 in Figure 6-15.

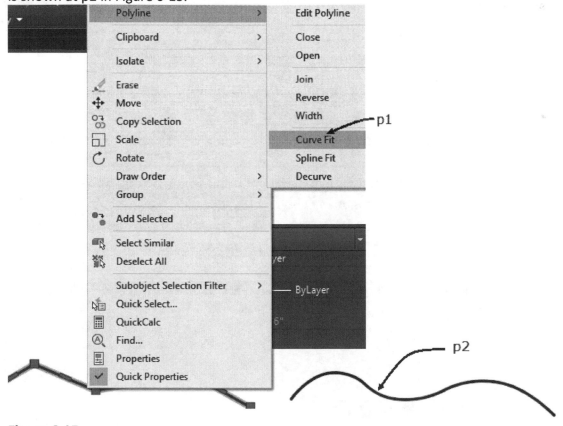

Figure 6-15

6.1 Using the Array Option of Copy

The following tutorial includes the use of the Array option in the Copy command. The Array option allows you to specify the number of items and either the distance between items or a distance to fit the number of items.

1. Open AutoCAD 2024 from the Desktop shortcut.
2. Choose **Open Files** from the **Quick Access Toolbar**.
3. Open the **Copy Array.dwg** drawing file from the **AutoCAD 2024 Certified User Exercise Files \ Exercise Files \ Ch 6** folder.
4. In the following steps, you will place lines for parking spaces using the Copy command.
5. Choose **Copy** from the Modify panel of the Home tab, as shown in Figure 6-16.

Figure 6-16

Respond to the workspace prompts as follows:

Command: _copy

Select objects: 1 found (Select the line shown at p1 in Figure 6-17.)

Select objects: **ENTER** (Press **ENTER** to end selection.)

Current settings: Copy mode = Multiple

Specify base point or [Displacement/mOde<Displacement>: (Select the end of the line at p1 as shown in Figure 6-17.)

Specify second point or [Array] <use first point as displacement>: **a** ENTER *(Press the down arrow key to toggle to the **Array** option; press **Enter** to choose the Array option.)*

Enter number of items to array: **40 ENTER** (Type **40**, press ENTER to specify 40 spaces.)

Specify second point or [Fit]: **11' ENTER** *(Type **11'** and press **Enter** to specify the distance between lines.)*

Specify second point or [Array/Exit/Undo] <Exit>: **ENTER** *(Press **ENTER** to exit the command.)*

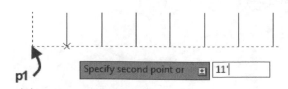

Figure 6-17

6. The 40 parking spaces are created as shown in Figure 6-18.

Figure 6-18

7. Choose the Annotate tab.
8. Choose the flyout shown at p1 in Figure 6-19 to display available styles. Select the Annotative style as shown in Figure 6-19.

Figure 6-19

9. The Annotative style is set current as shown in Figure 6-20.

Figure 6-20

10. To specify the Annotative Scale select the flyout shown at p1 in Figure 6-21 of the Status bar. Choose the **1/32" = 1'-0"**.

Figure 6-21

11. Toggle ON the *Add scales to annotative objects when the annotation scale changes* of the Status bar as shown in Figure 6-22.

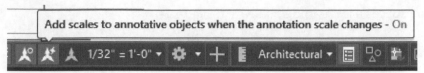

Figure 6-22

12. Choose **Multiline** text from the Text panel in the Annotate Tab.

    Specify first corner: (Select a point near p1 as shown in Figure 6-23.)

    Specify opposite corner or [Height/Justify/Line spacing/Rotation/Style/Width/Columns]: (Select a point near p2 as shown in Figure 6-23.)

    Type **1** in the on-screen text window. Choose **Close Text Editor** from the contextual tab.

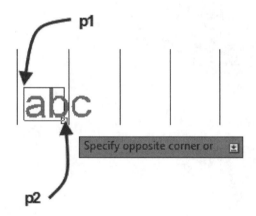

Figure 6-23

13. Choose **Copy** from the Modify panel in the Home tab. Copy the text placed in the previous step 5 times to the right using the Array option as shown in Figure 6-24.

Figure 6-24

14. Double click each number of text; edit the text as shown in Figure 6-25.

Figure 6-25

15. Choose **Tool Palettes** from the Palettes panel of View tab as shown in Figure 6-26.

Figure 6-26

16. Choose the Architectural palette. Choose **Trees-Imperial** block of the Architectural palette as shown in Figure 6-27.

Figure 6-27

17. Insert the **Trees-Imperial** block near parking slot 1 as shown in Figure 6-28.

Figure 6-28

18. Choose the **Copy** command from the Modify panel in the Home tab.

Command: _copy

Select objects: 1 found (Select the tree placed in the previous step.)

Select objects: (Press ENTER to end selection.)

Current settings: Copy mode = Multiple

Specify base point or [Displacement/mOde] <Displacement>: (Select a point near p1 using the Endpoint object snap as shown in Figure 6-29.)

Specify second point or [Array] <use first point as displacement>: **A** (Type A, press Enter to specify the Array option.)

Enter number of items to array: **5** (Type **5**, press **Enter** to specify the number of objects in the array.)

Specify second point or [Fit]: **F** (Type **F**, press **Enter** to specify the Fit option.)

Specify second point or [Array]: (Select a point using the Endpoint object snap mode at p2 as shown in Figure 6-29.)

Figure 6-29

19. The five trees are placed as shown in Figure 6-30.

Figure 6-30

20. Choose the Home tab. Expand the Utilities panel. Choose **Point** Style as shown in Figure 6-31.

Figure 6-31

21. Choose the point style from the Point Style dialog box as shown at **p1** in Figure 6-32.

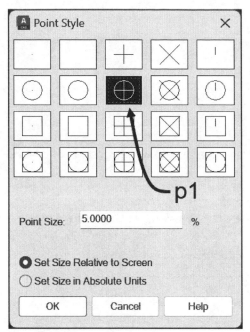

Figure 6-32

22. Use the Array option of the **Copy** command to develop a total of 30 parking slots 9' wide to the right of point p1 shown in Figure 6-33. Place 3 trees evenly spaced above the parking slots.

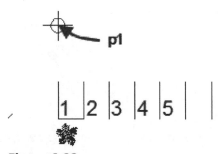

Figure 6-33

23. The parking slots are shown in Figure 6-34.

Figure 6-34

24. Save the drawing as **Copy Array Your Name** to your student folder and close the file.

6.2 Creating Rectangular Arrays

The following tutorial includes practice at creating a rectangular array.

1. Verify **Drafting** & **Annotation** is the current workspace as shown in the Quick Access toolbar.
2. Choose Open from the Quick Access toolbar.
3. Navigate to the **AutoCAD 2024 Certified User Exercise Files \ Exercise Files \ Ch 6** folder. Choose the **Array** drawing file.
4. Verify Dynamic Input is toggled on in the Status bar.
5. The geometry for an array is included in the Array drawing as shown in Figure 6-35.

Figure 6-35

6. Choose Rectangular Array from the Array flyout as shown at p1 in the Modify panel in the Home tab as shown in Figure 6-36.

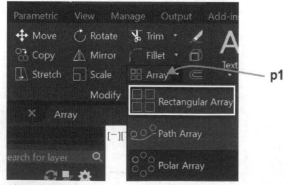

Figure 6-36

7. In the next series of steps you will create the array to place the circle in the rectangles as shown in Figure 6-37.

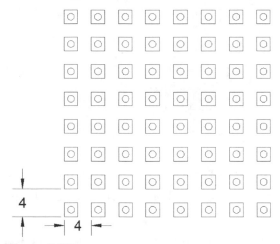

Figure 6-37

8. Select the circle shown at p1 in Figure 6-38.

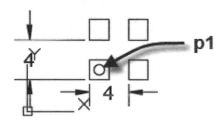

Figure 6-38

9. In the Array Creation tab, edit the number of columns to **8**, distance between columns to **4**, number of rows to **8**, and distance between rows to **4** as shown in Figure 6-39.

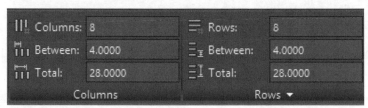

Figure 6-39

10. Choose **Close** Array of the Close panel in the Array Creation tab.
11. Save the file in your student directory. Name the file Rectangular Array_Your Name.

6.3 Creating Rectangular and Path Arrays

The following tutorial provides practice at creating rectangular arrays and the path array tool.

1. Open AutoCAD 2024 from the Desktop shortcut.
2. Choose **QNew** of the **Quick Access** toolbar as shown in Figure 6-40. Choose acad.dwt template of the Select template dialog box. Choose Open to dismiss the file dialog box.
3. Set **Drafting & Annotation** workspace current in the Quick Access toolbar.
4. Choose the Polygon command from the Draw panel. Respond to the Workspace prompts as shown below:
    Enter number of sides <4>: **6** (Type 6 press **ENTER** to specify the number of sides.)
    Specify center of polygon or [Edge]: **2,2** (Type **2,2** and press **ENTER** to specify the center of polygon location.)
    Enter an option [Inscribed in circle/Circumscribed about circle] <I>: **I** (Press **Enter** to choose the default Inscribed method.)
    Specify radius of circle: **.5** (Type **.5** press **ENTER** to specify the size of the circle.)

Figure 6-40

5. Choose **Rectangular Array** from the Modify panel.
    Respond to the command prompts as shown below.
    Select objects: 1 found (Select the hexagon created in the previous step.)
    Select objects: **Enter** *(Press **ENTER** to end object selection and open the Array Creation tab.)*
    *In the Properties panel verify Associative is **ON**. In Columns panel edit Columns to **5**, Between columns to **1** and in the Rows panel edit Rows to **3**, Between rows to **1** as shown in Figure 6-41.*
    *Choose **Close Array** of the Close panel in Array Creation tab.*

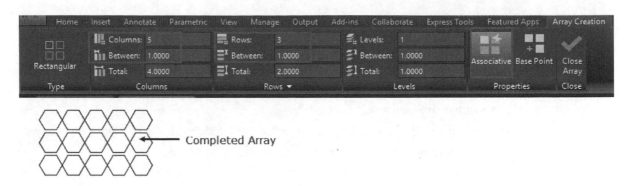

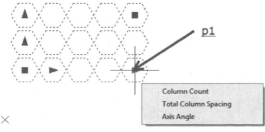

Figure 6-41

6. The array created in the previous step is associative. Select the array to display its grips. Hover over each grip to display the tips for the grips as shown in Figure 6-42.

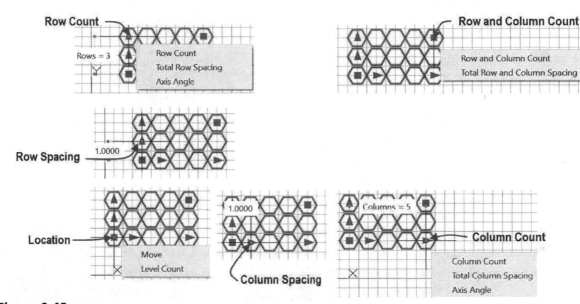

Figure 6-42

7. Select the grip as shown at p1 in Figure 6-43. Move the cursor to increase the column count to **6**. Press **Escape** to clear the grips.

Figure 6-43

8. Choose the **Polygon** command from the Draw toolbar. Respond to the Workspace prompts as shown below:

   Enter number of sides <4>: **6** (Type **6** and press **ENTER** to specify the number of sides.)

   Specify center of polygon or [Edge]: **9,2** (Type **9, 2** and press **ENTER** to specify the center location for the polygon.)

   Enter an option [Inscribed in circle/Circumscribed about circle] <I>: **I** (Type **I** and press **ENTER** to choose the Inscribed method of creating the polygon.)

   Specify radius of circle: **.5** (Type **.5** and press **ENTER** to specify the radius of the polygon.)

Figure 6-44

9. Choose the Rectangular Array command of the Modify panel.

   Respond to the workspace prompts as shown below.

   Select objects: 1 found (Select the hexagon created in the previous step.)

   Select objects: *(Press **ENTER** to end selection of objects.)*

   *In the Properties panel verify Associative is **OFF**. In Columns panel edit Columns to **3**, Between columns to **1** and in the Rows panel edit Rows to **3**, Between rows to **1** as shown in Figure 6-45.*

   *Choose **Close Array** of the Close panel in Array Creation tab.*

Figure 6-45

10. Select a polygon of the array created in the previous step to display its grips. Notice only the grips of a polygon are shown as displayed in Figure 6-46; therefore, the array settings cannot be edited for an array created without the associative property.

Figure 6-46

11. Press **Escape** to clear selection.

12. Note, when the ArrayRect command is selected in the future the Associative property defaults to No since you selected No while placing the last array.

13. Choose **SW Isometric** from the ViewCube as shown at **p1** in Figure 6-47.

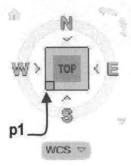

Figure 6-47

14. In this step you will create an additional level to the array you modified in step 7 to create a three-dimensional array. Select the array shown at **p1** in Figure 6-48 to open the Array tab. In the Levels panel edit the Levels = **2** and Between spacing to **5** as shown at p2 in Figure 6-48. Choose the **Close Array** of the Close panel to close the Array tab.

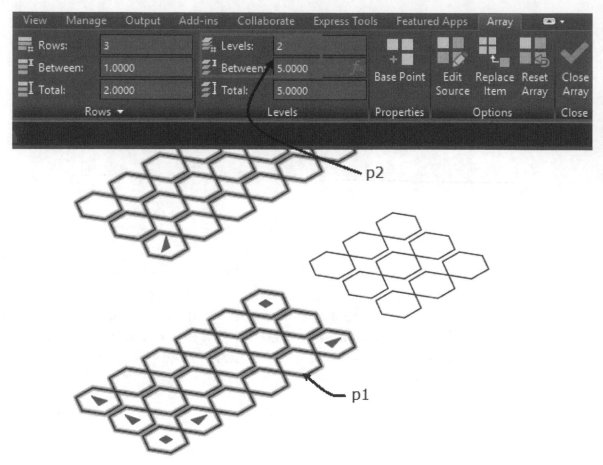

Figure 6-48

15. The 3D array is created as shown in Figure 6-49.

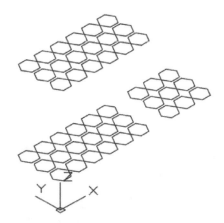

Figure 6-49

16. Choose **Top** from the ViewCube. In the next series of steps, you will create the arc which will be used for a path array.

17. Choose the Home tab. Choose **Line** from the Draw panel. Draw a **1"** line from the endpoint of the hexagon shown in Figure 6-50.

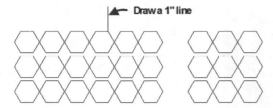

Figure 6-50

18. Choose **3 Point Arc** from the Draw panel. Draw an arc in a counterclockwise direction through points p1, p2, and p3 using the Endpoint object snap mode as shown in Figure 6-51.

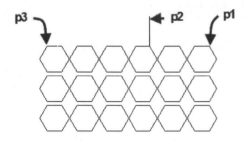

Figure 6-51

19. Choose the **Circle** command from the Draw panel. Create a circle centered at p1 with a radius of **.25** as shown in Figure 6-52.

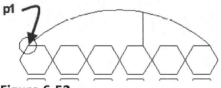

Figure 6-52

20. Choose **Path Array** from the Array flyout of the Modify panel as shown in Figure 6-53.

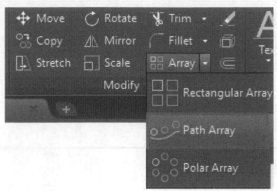

Figure 6-53

21. Respond to the command prompts as follows.

   Select objects: 1 found (Select the circle located at **p1** as shown in Figure 6-54.)

   Select objects: **Enter** (Press **ENTER** to end selection.)

   Select path curve: (Select the arc at **p2** in Figure 6-54.)

   Select grip to edit array or:

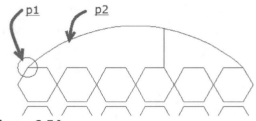

Figure 6-54

In the Properties panel in the Array Creation tab choose **Associative** toggle and choose **Divide** of the flyout as shown in the Properties panel at p1 in Figure 6-55.

Edit the number of Items to **5** in the Items panel of the Array Creation tab as shown at p2 shown in Figure 6-55.

Choose Close Array of the Close panel in the Array Creation tab.

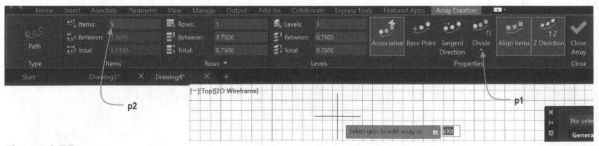

Figure 6-55

The array is created as shown in Figure 6-56.

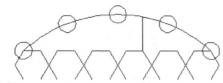

Figure 6-56

22. To demonstrate the associative property, you will edit the arc to change the distance between the circles, select a circle of the array shown at **p1** to display the Array contextual tab as shown in Figure 6-57. Note the Divide method is chosen in the Properties panel in the Array tab.

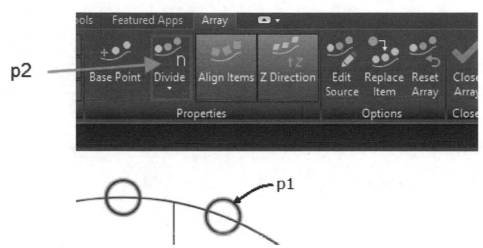

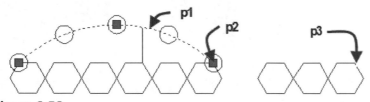

Figure 6-57

23. Select the arc at p1 as shown in Figure 6-58. Select the grip at p2, stretch the arc endpoint to p3 as shown in Figure 6-58.

Figure 6-58

The resulting array is shown in Figure 6-59.

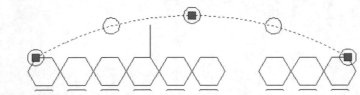

Figure 6-59

24. Choose **SaveAs** from the Quick Access Toolbar. Save the file as **Rectangular and Path Array _ your name** in your student directory.

25. Close the drawing.

6.4 Creating a Polar Array with Levels

The following tutorial provides practice creating polar arrays as shown in Figure 6-60.

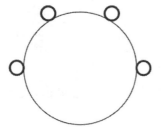

Figure 6-60

1. Open AutoCAD 2024 from the Desktop shortcut.
2. Choose **Open** from the **Quick Access** toolbar.
3. Choose the **Polar Array.dwg** file from the **AutoCAD 2024 Certified User Exercise Files \ Exercise Files \Ch 6** folder.
4. Choose **Polar Array** from the Modify panel in the Home tab.

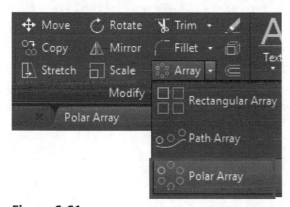

Figure 6-61

5. Respond to the workspace prompts as shown below.
   Select objects: 1 found *(Select the circle shown at p1 in Figure 6-62.)*
   Select objects: Press Enter to end selection.
   Specify center point of array or: *(Press Shift, right-click choose* Center, *move the cursor to p2 as shown in Figure 6-62 to display Center object snap marker; click to select the center.)*

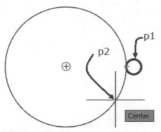

Figure 6-62

The Array Creation tab opens. Enter number of items to **4** and Fill to **180** in the Items panel as shown in Figure 6-63.

Figure 6-63

Verify **Associative** is toggled **ON** in the Properties panel. Choose **Close Array** to close the Array Creation tab.

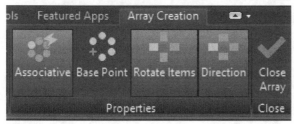

Figure 6-64

6. The completed array is shown in Figure 6-65.

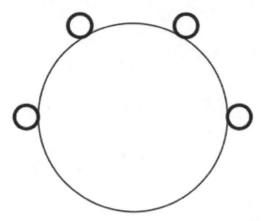

Figure 6-65

7. Refer to the presentation in the Polar Array and edit the array to a **120** degrees fill and **4** items as shown in Figure 6-66.

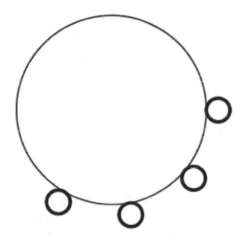

Figure 6-66

8. Save the drawing to your student folder and name the drawing Polar Array Your Name. Close the drawing.

6.5 Creating and Editing Polylines

The following tutorial provides practice in the creation and edit of polylines.

1. Open AutoCAD. Choose **QNew** from the **Quick Access** toolbar.
2. Verify acad.dwt is the selected template, choose Open to dismiss the **Select template** dialog box.
3. In the next series of steps, you will create an arrow using the polyline command.
4. Toggle ON Polar and Dynamic Input in the Status bar.

Figure 6-67

5. Choose the **Polyline** command from the Draw panel of the Home tab.

Figure 6-68

6. Respond to the workspace prompts as follows:
   Command: _pline
   Specify start point: **5,5** *(Press **Enter**)*
   (Press the **Down Arrow** key of the keyboard to choose the **Width** option.)
   Current line-width is 0.0000
   Specify next point or [Arc/Halfwidth/Length/Undo/Width]: **W**
   Specify starting width <0.0000>: **.25** (Type **.25** and press **Enter** to specify the starting width.)
   Specify ending width <0.2500>: **.25** (Type **.25** and press **Enter** to specify the ending width.)

   Move the cursor to the right to display a polar angle of **zero** as shown below.

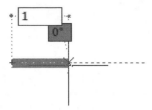

Figure 6-69

Specify next point or [Arc/Halfwidth/Length/Undo/Width]: **1** (Type **1** to create the first segment length.)
(Press the down arrow of the keyboard to choose the Width option.)
Specify next point or [Arc/Close/Halfwidth/Length/Undo/Width]: **w**
Specify starting width <0.2500>: **.75** (Enter **.75** and press **Enter** to change the start width.)
Specify ending width <0.7500>: **0** (Enter **0** and press **Enter** to change the ending width.)
Specify next point or [Arc/Close/Halfwidth/Length/Undo/Width]: **1** (Move the cursor to the right to display a polar angle of 0, type 1 press **Enter** in the workspace.)
Specify next point or [Arc/Close/Halfwidth/Length/Undo/Width (Press **Enter** to end the command; the arrow as shown below is created.)

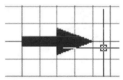

Figure 6-70

7. Choose limits command. Set the limits with lower left corner set to **0,0** and the upper right corner set to **24,12**.

8. Choose **Zoom All** from the Navigation bar. In the next step you will draw a polyline through the absolute coordinate points; therefore, toggle **OFF** Dynamic Input in the Status bar, since Dynamic Input uses Relative Polar.

9. Choose the Polyline command from the Draw panel of the Home tab. Refer to Figure 6-71; draw a polyline with zero width through the following points:

   5, 7

   8, 8

   12, 7

   14, 5

   15, 7

   19, 8

   22, 7

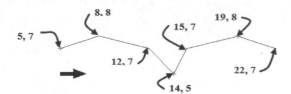

Figure 6-71

10. Select the polyline and copy it using the Endpoint object snap mode at **p1** as the basepoint to the coordinate **5, 10** as shown in Figure 6-72.

Figure 6-72

11. Select the polyline at p1 as shown in Figure 6-73, and right click choose **Polyline > Curve Fit** from the contextual menu as shown below.

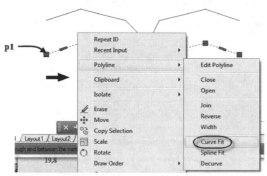

Figure 6-73

12. The polyline is curved to pass through the vertex points of polyline as shown in Figure 6-74.

Figure 6-74

13. Select the polyline at p1 as shown in Figure 6-75; right click choose **Polyline > Spline Fit** from the contextual menu.

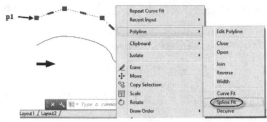

Figure 6-75

14. Note the polyline creates a curve to fit using the vertices as control points as shown in Figure 6-76.

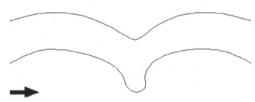

Figure 6-76

15. Choose **Save** from the Quick Access toolbar. Save the drawing as **Polyline_Your_Name** to your student folder. Close the file.

6.6 Exploring Hatch Options, Boundary Enhancements and Associative Hatch

The following tutorial provides practice in placing hatch patterns using island detection methods, associative hatch, annotative hatch and boundary detection error tool.

1. Open the **Hatch.dwg** drawing file from the **AutoCAD 2024 Certified User Exercise Files \ Exercise Files \ Ch 6** folder.
2. Verify **Quick Properties** is toggled ON in the Status bar.
3. Set the **Hatch** layer current. Choose the Normal layout.
4. In the next series of steps, you will place a hatch using the Normal Island Detection. The Normal Island Detection will place hatch patterns **inward** from the outer boundary. When an internal island is encountered hatching is turned off until another island is encountered.
5. Choose **Hatch** from the flyout shown at p1 in Figure 6-77 in the Draw panel of the Home tab.

Figure 6-77

6. Verify the **ANSI 31** pattern is the current pattern as shown in Pattern panel of the Hatch Creation tab in Figure 6-78.

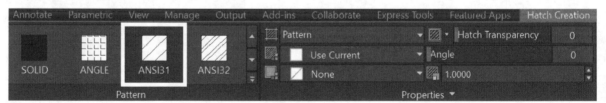

Figure 6-78

7. As shown in Figure 6-79 toggle off Associative Hatch at p1 and extend the Options panel as shown at p2. Choose the **Island Detection** flyout at p3 to display the detection options and choose **Normal Island Detection** method from the flyout shown at p4 in Figure 6-79.

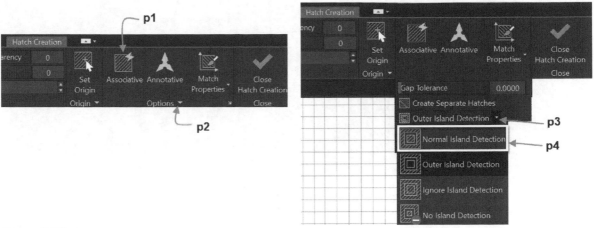

Figure 6-79

8. Choose the **Select** button of the Boundaries panel as shown in Figure 6-80.

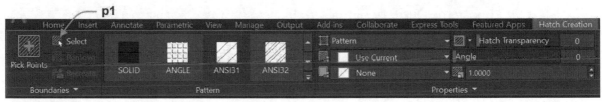

Figure 6-80

9. Respond to the workspace prompts as follows:

Select objects or [picK internal point/Undo/seTtings]: (Left click at **p1** as shown in Figure 6-81 and release the left mouse button.)

Specify opposite corner: (Move the cursor to **p2** as shown in Figure 6-81 and click to end selection.) 4 found

Select objects or [picK internal point/Undo/seTtings]: (Press **Enter** to end selection.)

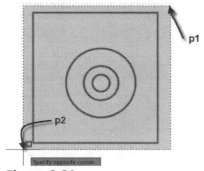

Figure 6-81

10. The Hatch created using the Normal setting is shown in Figure 6-82.

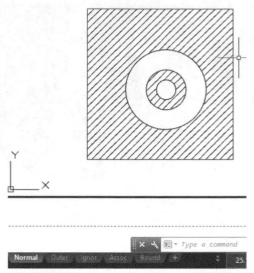

Figure 6-82

11. Choose the **Outer** layout tab. In the next step, you will place a hatch using the Outer Island Detection. The Outer option is recommended. The hatch pattern is placed **inward** from the outer boundary and leaves the internal islands unaffected.

12. Repeat steps 5-9, edit the Island Detection to **Outer** to place the hatch pattern as shown in Figure 6-83.

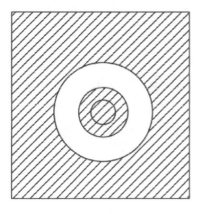

Figure 6-83

13. Choose the Ignor layout tab. In this step, you will place a hatch using the Ignor Island Detection method. Repeat steps 5-9 and edit the Island Detection to Ignor to place the hatch pattern as shown in Figure 6-84.

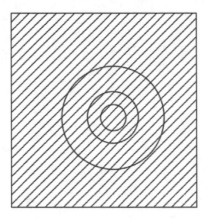

Figure 6-84

14. Select the layout flyout at p1 as shown in Figure 6-85; choose the Assoc layout.

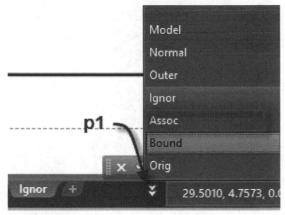

Figure 6-85

15. In the next series of steps, you will place an associative hatch. The associative hatch edits to fill the boundary when the boundary geometry changes.

16. Choose **Hatch** from the **Draw** panel of the Home tab.

17. Verify **ANSI 31** is the current pattern from the Pattern panel of the **Hatch Creation** tab. In the Options panel, choose the Associative button as shown in Figure 6-86 below. Verify the Normal Island Detection is current.

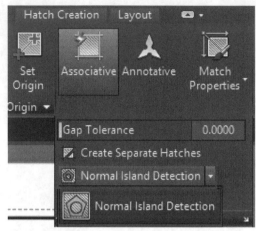

Figure 6-86

18. Refer to Figure 6-87; respond to the workspace prompts as follows:

Select objects or [picK internal point/Undo/seTtings]: (Left click at p1; release the left mouse button.)

Specify opposite corner: (Move the cursor to p2 and click to end selection.) 4 found

Select objects or [picK internal point/Undo/seTtings]: (Press Enter to end selection.)

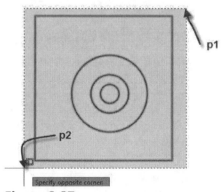

Figure 6-87

19. The hatch pattern is placed as shown in Figure 6-88.

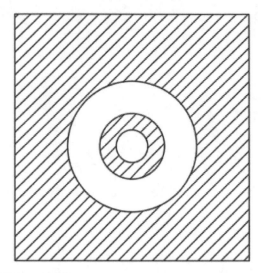

Figure 6-88

20. Select the rectangle as shown at p1 in Figure 6-89. Caution: do not select the hatch pattern; the polyline entity type is displayed in the Quick Properties palette.

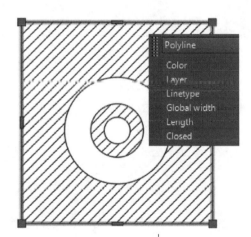

Figure 6-89

21. Select the grip as shown at p1 and stretch the polyline down as shown in Figure 6-90.

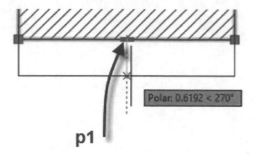

Figure 6-90

22. Note the hatch pattern is associative since it expands to fill the boundary as shown in Figure 6-91.

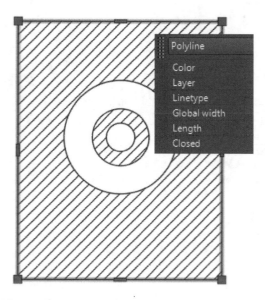

Figure 6-91

23. In the next series of steps, you will place a hatch within a boundary that consists of gaps and experience the boundary detection tools. Choose the Bound layout from the layout flyout list.

24. Choose **Hatch** from the **Draw** panel of the Home tab.

25. Verify **ANSI 31** is the current pattern from the Pattern panel of the **Hatch Creation** tab shown in Figure 6-92.

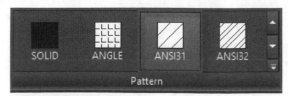

Figure 6-92

26. Verify **Associative** is current of the Options panel in the **Hatch Creation** tab.

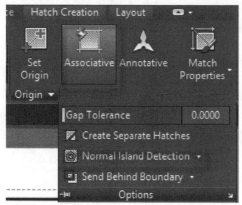

Figure 6-93

27. Choose **Pick Points** from the Boundaries panel of the **Hatch Creation** tab.
28. Move the cursor to the workspace; choose a point near **p1** as shown in Figure 6-94.

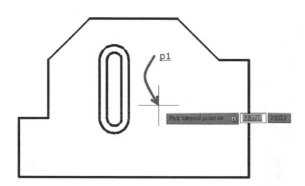

Figure 6-94

29. The gap in the boundary of the shape is identified as shown in Figure 6-95. Choose Close to accept the message.

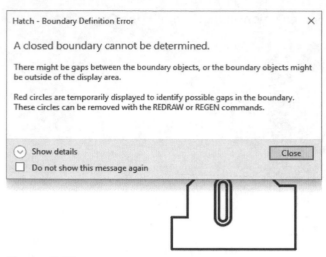

Figure 6-95

30. Expand the **Options** panel of the Hatch Creation tab to edit the gap tolerance as shown in Figure 6-96.

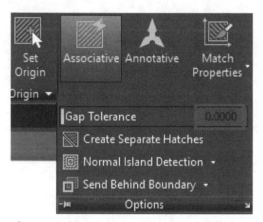

Figure 6-96

31. In this step you edit the tolerance allowing the placement of the hatch. Edit the Tolerance to **.125** in the Gap tolerance section as shown in Figure 6-97.

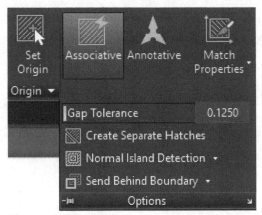

Figure 6-97

32. Choose **Pick Points** of Boundaries panel; click a point at p1 in Figure 6-98.

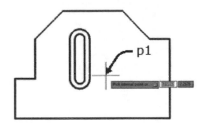

Figure 6-98

33. Since the gap in the boundary exists the following Hatch – Open Boundary Warning
    message box will open to alert you of the problem as shown in Figure 6-99. You can
    choose the **Continue hatching this area** or choose the **Do not hatch this area** text and
    either fix the boundary or you can increase the gap tolerance to place the hatch.

  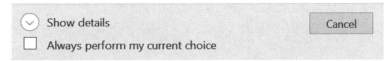

Hatch - Open Boundary Warning   ✕

The hatch boundary is not closed. What do
you want to do?

→ Continue hatching this area
The area will be hatched even though one or more gaps
exist.

→ Do not hatch this area

Show details   Cancel
☐ Always perform my current choice

Figure 6-99

34. If you choose the **Continue hatching this area** text choose **Close Hatch Creation** from the Close panel as shown in Figure 6-100 at right of the Hatch Creation tab.

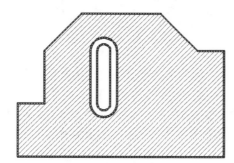

Figure 6-100

35. The hatch is placed as shown in Figure 6-101.

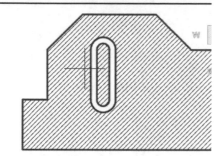

Figure 6-101

36. Select the hatch to display the Hatch Editor tab.
37. Choose the **ANSI 32** to change the pattern from the Pattern panel shown in Figure 6-102.

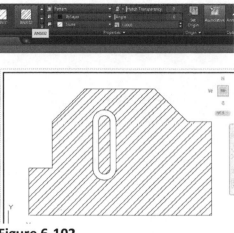

Figure 6-102

38. Move the cursor over the grip of the hatch, notice the options in the tip for the edit of the hatch in Figure 6-103.

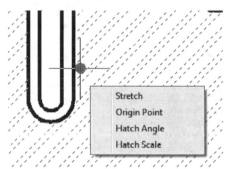

Figure 6-103

39. Choose the Hatch Angle option, move the cursor down to display a polar angle of **270** as shown in Figure 6-104, left click to set the angle.

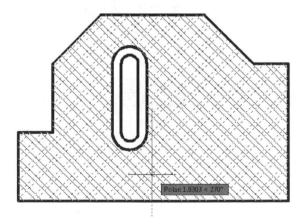

Figure 6-104

40. Choose the **Close Hatch Editor** command of the Close panel in the Hatch Editor tab. The completed hatch edit is shown in Figure 6-105.

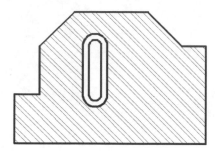

Figure 6-105

41. In the next series of steps, you will edit the hatch pattern to annotative which controls the hatch scale to match the viewport scale. Select the hatch shown in Figure 6-105 and choose **Annotative** in the Options panel as shown in Figure 6-106.

Figure 6-106

42. In the previous steps you created an Annotative hatch; the scale of this pattern will adjust to the viewport scale. In the final steps you will view the scale of the hatch pattern as the viewport scale changes.

43. Toggle ON *Add Scales to annotative objects when the annotation scale changes* as shown in Figure 6-107.

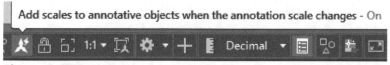

Figure 6-107

44. Double click inside the viewport to make it active as shown in Figure 6-108.

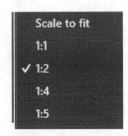

Figure 6-108

45. Choose the **1:2** scale from the Viewport scale flyout as shown in Figure 6-109.

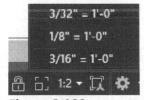

Figure 6-109

46. The part and the hatch are scaled as shown in Figure 6-110.

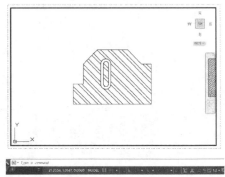

Figure 6-110

47. Choose the **1:4** scale from the Viewport Scale flyout.

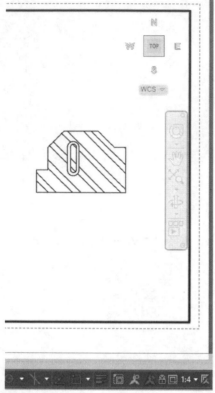

Figure 6-111

48. Note the lines representing the hatch retain the same scale although the overall part size has changed to a different scale.

49. In the next series of steps, you will use the Set Origin tool to change the origin of the hatch pattern.

50. Choose the **Orig** layout tab from the layout flyout list as shown in Figure 6-112.

Figure 6-112

51. Select the hatch pattern to open the Hatch Editor. Note the Pattern name is Brick and the scale is 24 as shown in Figure 6-113.

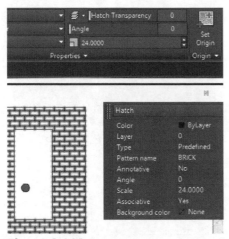

Figure 6-113

52. Choose the **Set Origin** button of the Origin panel as shown in Figure 6-114.

Figure 6-114

53. Respond to the workspace prompts as follows:

Move the cursor to p1 and click as shown in Figure 6-115.

Choose the Close Hatch Editor to end the edit as shown at p2 in Figure 6-115.

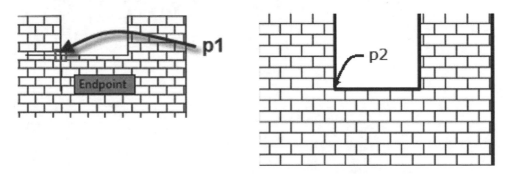

Figure 6-115

54. Save the drawing as **Hatch_Your_Name** in your student folder and close the drawing.

Notes:

Chapter 7
Annotating Drawings

This chapter will review the methods of placing and editing single line and multiline text. The chapter also includes a review of dimensioning commands. There are three tutorials and a text justification quiz at the end of the chapter.

The student will be able to:

1. Create and edit single line and multi-line text to a specific annotative scale
2. Create annotative scales for text, dimensions and multileaders
3. Remove annotative scale representations that are not needed in a drawing
4. Insert special characters and symbols in single line and multiline text
5. Insert and edit center marks

Single Line Text

The Single line text command is located on the Annotation panel in the Home tab and on Text panel in the Annotate tab. Choose the flyout shown at p1 to view the Multiline or Single Line text commands as shown at p2 in Figure 7-1.

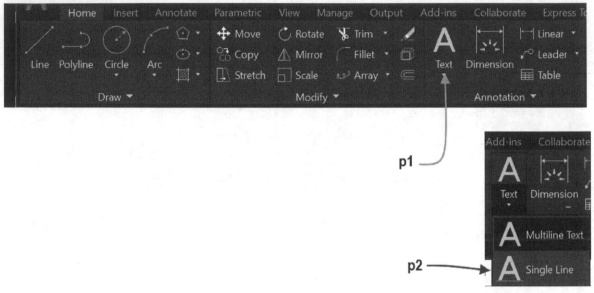

Figure 7-1

When you choose the Single Line text command, you are prompted to specify a start point for single line text placement, as shown at p1 in Figure 7-2. However, you can press the down arrow of the keyboard to change the text justification or text style. The justify options are shown in Figure 7-3 for your review. The test could ask you to identify the justify setting by display of grips.

Notice you are prompted at p2 to specify the height of text. If you move the cursor and click you will specify the height. However, you can insert the height of text by typing the number value in the workspace prompt. Finally, you are prompted to specify the rotation as shown at p3 in Figure 7-2. After entering the rotation you can begin typing text. You can press enter to place multiple lines of text. At any point when you are typing you can click to specify a different location using the same style, height and rotation. When you finish typing text press Enter twice to exit the command.

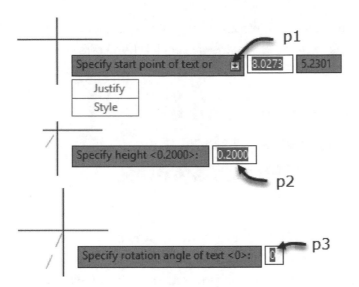

Figure 7-2

The Justify settings shown in Figure 7-3 of the Quick Properties palette include the fifteen justify settings available. Note that Center is the mid-length of the text and Middle is located at the mid-height and mid length of the Single Line text shown at left in Figure 7-3.

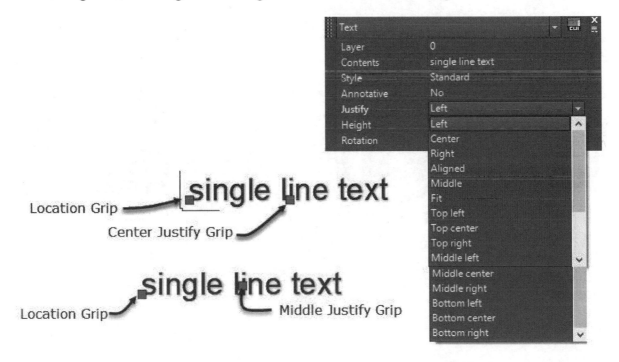

Figure 7-3

To edit existing text double click the text and add or delete characters at the cursor. To insert special characters such as the degree symbol type %%d. Additional special characters are shown in Figure 7-4.

Ø **Type %%C for diameter symbol**
° **Type %%D for degree symbol**
± **Type %%P for Plus / Minus symbol**

Figure 7-4

The text style supporting the Single line text and Multiline text can be viewed in the extended Annotation panel of the Home tab if you click the Text Style button shown at p1 to open the Text Style dialog box shown at left in Figure 7-5. You can change the current text style or view available styles from the flyout shown at p2 in Figure 7-5. To change the Font used in a style click the flyout shown at p3 in Figure 7-5. Click the New button shown at p4 to create a new text style as shown in Figure 7-5. Click at p5 of Annotative check box to create a text style which scales the text height by the Annotative scale and Viewport scale setting in the Status bar.

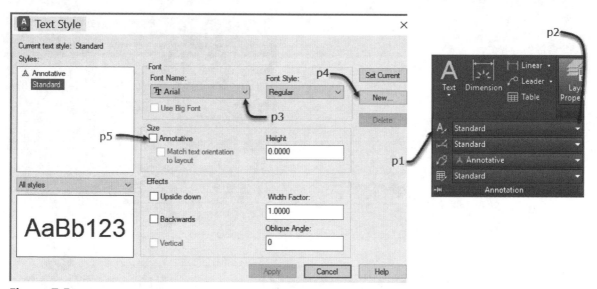

Figure 7-5

Multiline Text

Multiline text can be accessed from the Text panel in the Annotate tab as shown in Figure 7-6. The panel includes Spell Check shown at p1 and a flyout of text styles shown at p2. The flyout shown at p3 allows you to choose either Multi-line text or Single line text. The arrow shown at p4 shown at Figure 7-6 opens the Text Style dialog box shown in Figure 7-5.

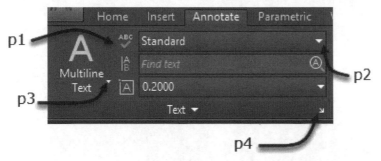

Figure 7-6

Multiline text is used to create text formatted as paragraphs; therefore, when you start the command you are prompted to Specify first corner as shown at p1 in Figure 7-7. After clicking at p2 move the cursor to p3 and specify opposite corner for text placement. The Text Editor opens as shown in Figure 7-7 and you can begin typing in the workspace below the ruler shown at p4.

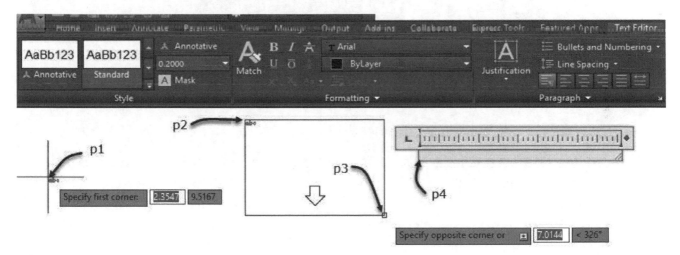

Figure 7-7

If Spell Check is turned on as shown at p1 a misspelled word will be underlined as shown at p2 in Figure 7-8. To correct the error move the cursor over the word and right-click; the contextual menu includes suggested word spelling which you can select as shown at p3 in Figure 7-8.

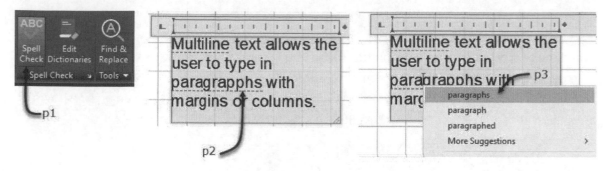

Figure 7-8

To insert special symbols in multi-line text, choose the Symbol flyout in the Insert panel of the Text Editor shown at p1 of Figure 7-9. The special symbols are shown at p2 of Figure 7-9.

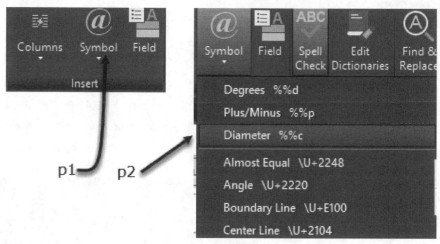

Figure 7-9

Creating Column Properties

When you insert text in the Text Editor do not specify column setting just start typing. After entering the text, select the text as shown at p1 in Figure 7-10. From the Insert panel of the Text Editor click Columns to choose Dynamic Columns > Manual height as shown at p2 in Figure 7-10. The text is resized as shown at p3 in Figure 7-10.

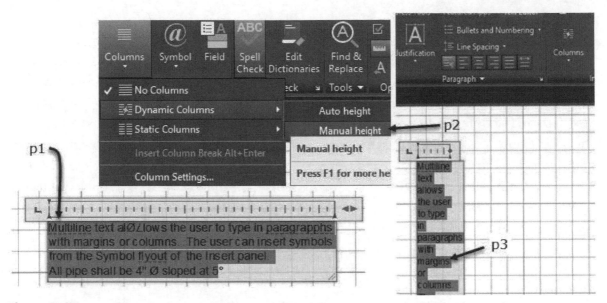

Figure 7-10

Retain the selection of the text.

From the Insert panel choose Columns and select Column Settings as shown at p1 in Figure 7-11. Edit the Column Settings dialog box Width section to Column=2 and Gutter=.5 as shown at p2 in Figure 7-11. Click near the lower boundary of the text at p3 in Figure 7-11 and drag up to create two columns as shown at p4.

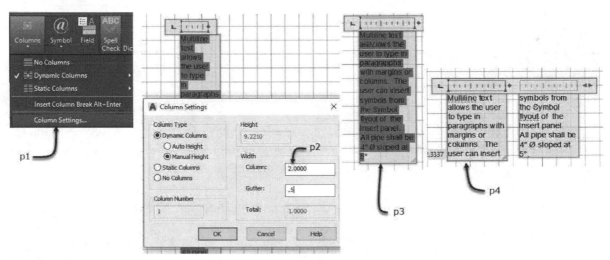

Figure 7-11

Dimensioning

The tools used for dimensioning are located on the Dimensions, Centerlines, and Leaders panels in the Annotate tab as shown in Figure 7-12. The Dimensions panel includes a flyout list of dimensions styles in the current drawing (shown at p1 in Figure 7-12) which includes the Standard and Annotative styles. Click the arrow at p2 to open the Dimension Style Manager as shown in Figure 7-13. The flyout at p3 includes the major dimension commands as shown at left in Figure 7-12. The QDim command shown at p4 in Figure 7-12 prompts you to select an object and the software determines the tool for dimensioning the object. The Dim command shown at p5 allows you to select or hover over an entity and the software will show a preview of the suitable dimensioning tool; just click to start the dimension. The QDim and Dim commands save time since you do not need to select from the flyout at p3 in Figure 7-12.

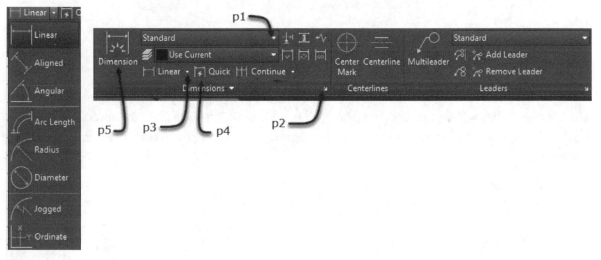

Figure 7-12

The Centerlines panel includes the Center Mark and Centerlines commands. When you select the Center Mark command from the Centerlines panel in the Annotate tab as shown at p1 in Figure 7-13 you are prompted as follows:

Select circle or arc to add center mark: (Select the circle as shown at **p2** in Figure 7-13.)

Select circle or arc to add center mark: (Press **Enter** to end select and view the center mark shown at p3 in Figure 7-13.)

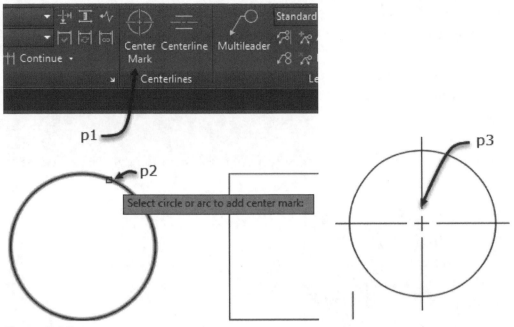

Figure 7-13

If you toggle on Quick Properties and select a Center Mark, you can view the settings for the extension as shown at p1 in Figure 7-14. When you toggle to NO for the Show extension the Center Mark is changed to two short lines as shown at p2 in Figure 7-14. To adjust the distance the center mark extends beyond the circle edit the extensions value as shown at p3 in Figure 7-14.

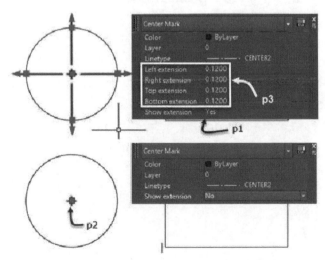

Figure 7-14

7.1 Inserting Single Line and Multiline Text

The following tutorial includes an exercise in placing Single Line and Multiline text.

1. Verify **Drafting** & **Annotation** is the current workspace as shown in the Quick Access toolbar.
2. Choose **Open** from the Quick Access toolbar. Navigate to the **AutoCAD 2024 Certified User Exercise Files \ Exercise Files \ Ch7** folder and choose the **Text 1.dwg** file. Save the file as Text 1_ Your Name
3. Choose the Home tab.
4. Choose the Single Line Text from the Text flyout of the Annotation panel in the Home tab as shown in Figure 7-15.

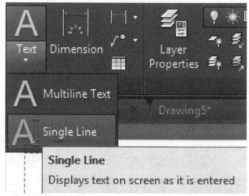

Figure 7-15

5. Respond to the workspace prompts as follows:
   Specify start point of text or [Justify/Style]: **J** (Press the **Down** arrow of the keyboard. Choose the **Justify** option.)
   Enter an option [Align/Fit/Center/Middle/Right/TL/TC/TR/ML/MC/MR/BL/BC/BR]: **c** (Press the **Down** arrow of the keyboard. Choose the **Center** option.)
   Specify center point of text: **6,0** (Type **6,0** press **Enter** to specify the center point.)
   Specify height <0.2000>: **.25** (Type **.25** to change the text height.)
   Specify rotation angle of text <0>: (Press **Enter** to accept the default rotation.)
   (Type **BRICK** in the workspace; press **Enter** twice to end the command.)
   Command:
6. Repeat the Single Line text command and place text Center justified at 6, 0.25 and 1/8" high for **SCALE = 1:1** as shown in Figure 7-16.

BRICK

SCALE = 1:1

Figure 7-16

7. Choose the Home tab, expand the Annotation panel, choose Text style as shown at p1 in Figure 7-17.

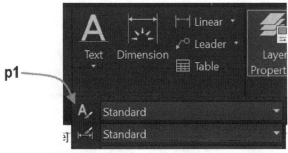

Figure 7-17

8. When the Text Style dialog box opens as shown in Figure 7-18 answer the following questions:

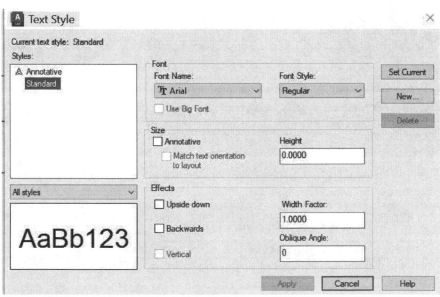

Figure 7-18

Note below the current Font Name and the height of text.

_____Font Name

_____Text Style Height

9. Is the current style Annotative?

10. Choose the Annotate tab. Choose the Text Style flyout shown at p1 in Figure 7-19 from the Text panel; select Annotative style as shown at p2.

Figure 7-19

11. In the next series of steps you will place annotative text which will retain the text height of 0.125 regardless of plot scale. Choose the Multiline text command from the Text panel in the Annotate tab. Verify the text height is .125 in the Text panel.

    Move the cursor to the workspace. Choose the 6" = 1'-0" from the scale flyout shown in Figure 7-20.

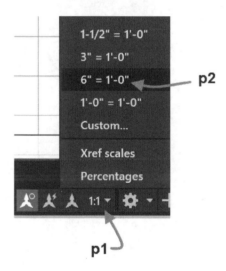

Figure 7-20

12. Respond to the workspace prompts as follows:

    Specify first corner: Select a point near p1 as shown in Figure 7-21.
    Specify opposite corner: Select a point near p2 as shown in Figure 7-21.

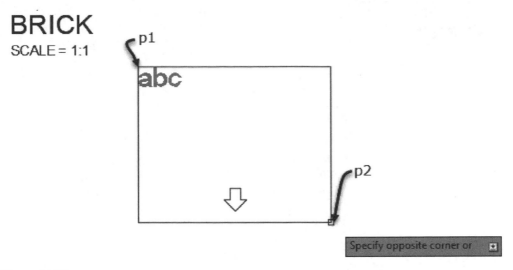

Figure 7-21

Type the text shown in Figure 7-22. The words Material Data should be bold format. Choose Close Text Editor to dismiss the Text Editor.

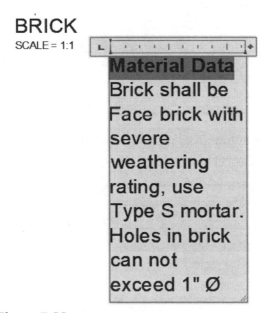

Figure 7-22

13. View the text height as shown in Figure 7-23. Note any future changes of the text at a different scale will result in the plot height of 0.125.

# BRICK
SCALE = 1:1

Material Data
Brick shall be
Face brick with
severe
weathering
rating, use
Type S mortar.
Holes in brick
can not
exceed 1" Ø

Figure 7-23

14. Save and close the drawing.

7.2 Inserting Basic Dimensions and Styles

The following tutorial includes placing Dimensions and creating a Dimension Style. The Geometric Center object snap mode is used in this tutorial.

1. Verify **Drafting** & **Annotation** is the current workspace in the Quick Access toolbar.
2. Choose **QNew** from the Quick Access toolbar.
3. Using the Rectang command draw a rectangle 7.625 x 3.625 with the lower left corner positioned at 2,2 as shown in Figure 7-24.

Figure 7-24

4. Toggle ON Dynamic Input in the Status bar.
5. Click the Object Snap toggle flyout of the Status bar, choose Geometric Center, Endpoint, Extension, Midpoint and Center object snap modes as shown in Figure 7-25.

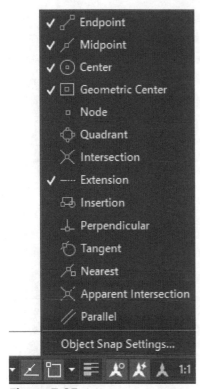

Figure 7-25

6. Choose the Object Snap Settings from Object Snap flyout of the Status bar to open the Drafting Settings dialog box. Verify Object Snap and Object Snap Tracking are toggled ON in the Drafting Settings dialog box as shown in Figure 7-26. Choose OK to dismiss the dialog box.

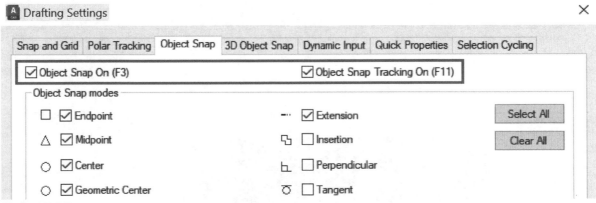

Figure 7-26

7. Choose the Circle command of the Draw panel in the Home tab.
8. Respond to the command prompts as follows:

Specify center point for circle or [3P/2P/Ttr (tan tan radius)]: (Hover the cursor over the rectangle and move the cursor to the center of the rectangle **p1** as shown in Figure 7-27 to display the **Geometric Center** object snap marker. Left click to specify the center.) Specify radius of circle or [Diameter] <0.5000>: **.5** (Move the cursor to dynamically display the circle; type **.5** press **Enter** to specify the radius.)

Figure 7-27

9. Use the Copy command to copy the circle **2"** to the left and right as shown in Figure 7-28.

Figure 7-28

10. Use the **Copy** command and copy the rectangle and the three circles 10" to the right as shown in Figure 7-29.

Figure 7-29

11. Choose the Annotate tab.
12. Choose the Linear Dimension from the Dimension flyout as shown in Figure 7-30.

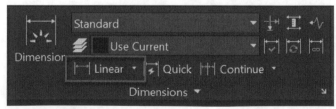

Figure 7-30

13. Respond to workspace prompts as shown below.

    Command: _dimlinear

    Specify first extension line origin or <select object>: (Select the endpoint of the horizontal line at **p1** as shown in Figure 7-31.)

    Specify second extension line origin: (Select the endpoint of the horizontal line at **p2** as shown in Figure 7-31.)

    Specify dimension line location or

    [Mtext/Text/Angle/Horizontal/Vertical/Rotated]: (Select a point near **p3** as shown in Figure 7-31.)

    Dimension text = 7.6250.

Figure 7-31

14. Add additional dimensions as shown in Figure 7-32.

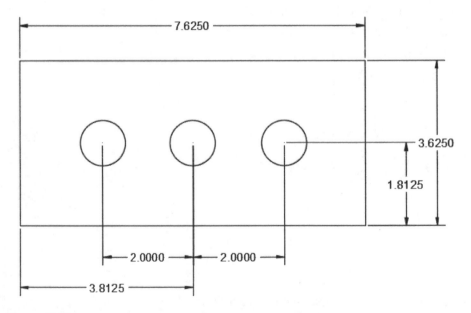

Figure 7-32

15. Choose the Diameter dimension as shown in Figure 7-33 from the Dimension flyout of the Dimensions panel.

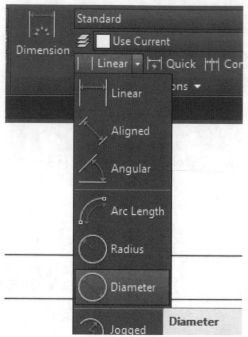

Figure 7-33

16. Respond to the workspace prompts as follows:

    Select arc or circle: (Select the circle at p1 as shown in Figure 7-34.)

    Dimension text = 1.0000

    Specify dimension line location or [Mtext/Text/Angle]: (Select a location near p2 as shown in Figure 7-34.)

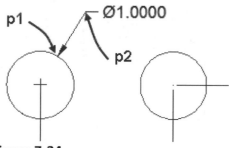

Figure 7-34

17. In the next series of steps, you will modify a dimension.
18. Toggle ON Quick Properties in the Status bar.
19. Select the text placed in the previous step to display its grips and Quick Properties palette as shown below.

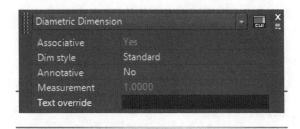

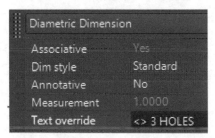

Figure 7-35

20. To add text following the diameter dimension type **<> 3 HOLES** in the Text override as shown in Figure 7-36. Note the <> characters retain the automatic update of the dimension if the circle diameter changes.

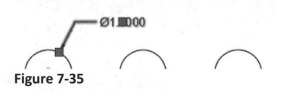

Figure 7-36

21. In the next series of steps, you will create and apply a dimension style to the existing dimensions.

22. Choose the Launching Dialog arrow shown at p1 in Figure 7-37 to open the Dimension Style Manager.

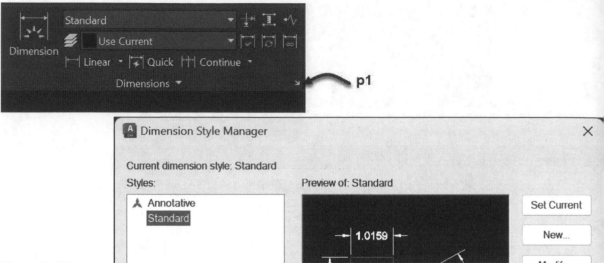

Figure 7-37

23. Verify Standard style is selected. Choose New as shown at right in Figure 7-38.

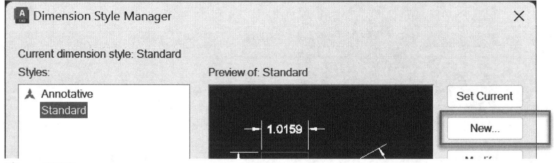

Figure 7-38

24. Overtype **ARC Detail** to change the new style name from Copy of Standard as shown in Figure 7-39. Press Continue to dismiss the Create New Dimension Style dialog box.

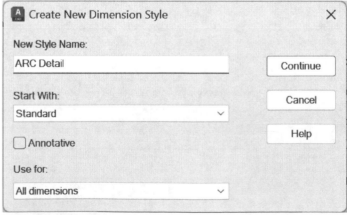

Figure 7-39

25. Choose the Primary Units tab of the New Dimension Style dialog box as shown in Figure 7-40. Choose Fractional from the Unit format flyout.

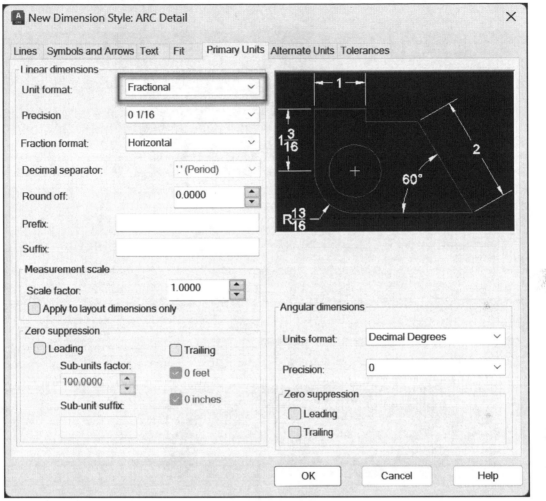

Figure 7-40

26. Choose OK to dismiss the dialog box.
27. Verify the ARC Detail style is selected as shown at left in Figure 7-41. Choose **Set Current** as shown at right in Figure 7-41.

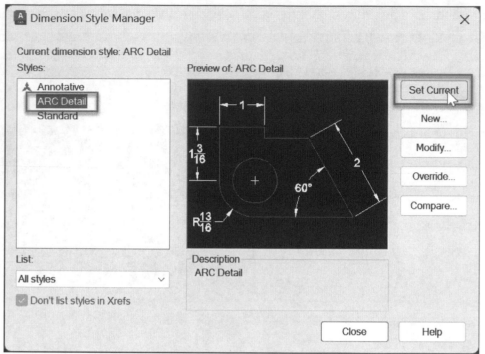

Figure 7-41

28. Choose **Close** to dismiss the dialog box.
29. Choose the **Update** tool as shown at p1 in Figure 7-42.

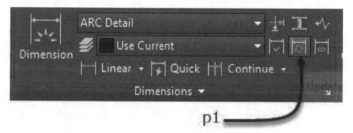

Figure 7-42

30. Respond to the workspace prompt as shown below.

    Select objects: all (Type **ALL** and press Enter to select all dimensions)
31. The dimensions are revised as shown in Figure 7-43.

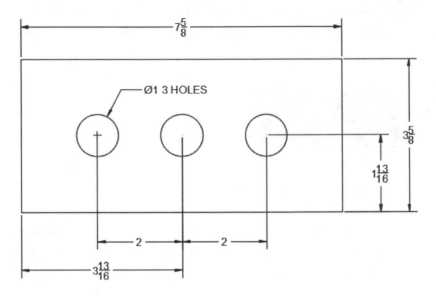

Figure 7-43

32. To add center lines for the circles you will use the Center Mark command of Centerlines panel in the **Annotate** tab shown in Figure 7-44.

Figure 7-44

33. Respond to the workspace prompts as shown below.

Select arc or circle: (Select the circle shown at **p1** in Figure 7-45.)
Select arc or circle: (Select the circle shown at **p2** in Figure 7-45.)
Select arc or circle: (Select the circle shown at **p3** in Figure 7-45.)

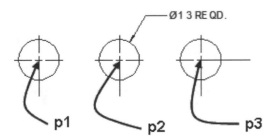

Figure 7-45

34. In the next series of steps, you will apply the DIM command to place dimensions. The DIM command allows you to place dimensions by selecting the object. Choose the **DIM** command as shown in Figure 7-46.

Figure 7-46

Respond to the workspace prompts as follows:

Select objects or specify first extension line origin or [Angular/Baseline/Continue/Ordinate/aliGn/Distribute/Layer/Undo]: (**Hover** the cursor over the line shown at p1 in Figure 7-47 to display the dimension. **Left click** the line and click above the dimension line to place the dimension.)

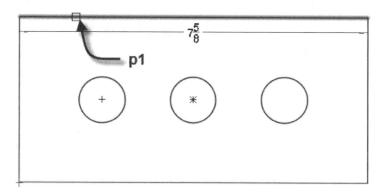

Figure 7-47

Select objects or specify first extension line origin or [Angular/Baseline/Continue/Ordinate/aliGn/Distribute/Layer/Undo]: (**Hover** the cursor over the line shown at p2 in Figure 7-48 to display the dimension. **Select** the line and left click at p3 to place the dimension line.)

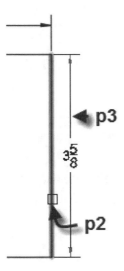

Figure 7-48

Select objects or specify first extension line origin or
[Angular/Baseline/Continue/Ordinate/aliGn/Distribute/Layer/Undo]: (**Hover** the cursor
over the circle shown at p4 in Figure 7-49 to display the dimension. **Select** the line and
left click at p5 to place the leader line.)

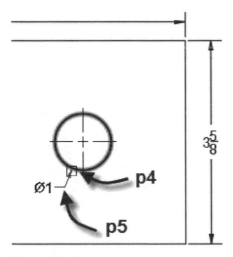

Figure 7-49

35. Save the drawing as Basic Dim _ Your Name in your student directory and close the
drawing.

7.3 Using Annotative Dimensions and Multileaders

The following tutorial provides an introduction to the use of annotative dimensions and multileaders.

1. Launch AutoCAD 2024.
2. Choose from the Application Menu, New > Drawing to open the **Select template** dialog box. Choose acad.dwt. Choose **Open** to dismiss the dialog box.
3. Verify the **Drafting & Annotation** is the current workspace in the Quick Access toolbar.
4. Select the **Home** tab. Choose **Layer Properties** as shown in 7-50 to open the Layer Properties Manager.

Figure 7-50

5. Click New **Layer** as shown in Figure 7-51. Create layers named **Object** and **Dimension**. Double click on the **Object** layer to make it current as shown at p2 in Figure 7-51.

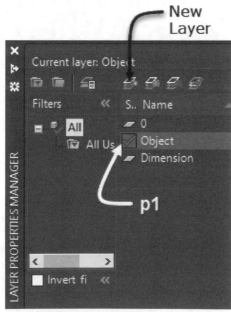

Figure 7-51

6. Choose the **Rectang** command, draw a rectangle **36"** wide and **80"** high with the start point at **2,2** as shown in Figure 7-52.

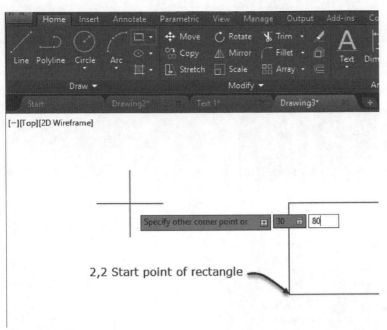

Figure 7-52

7. Choose **Zoom All** from the Navigation bar.

8. Roll the mouse wheel back to reduce the magnification.

9. Toggle Off the Grid in the Status bar to view the rectangle as shown in Figure 7-53.

Figure 7-53

10. Set the **Dimensions** layer current in the Layers panel.

11. In the next series of steps, you will create a dimension style for architectural dimensioning. Choose the Annotate tab. Choose the **Annotative** dimension style from the Dimensions panel as shown in Figure 7-54.

Figure 7-54

12. Click the arrow in the lower right corner of the panel to open the Dimension Style Manager as shown in Figure 7-55.

Figure 7-55

13. Click **New** shown at p1 in Figure 7-56 to create a new dimension style from the Annotative style. Type **Architectural** in the Create New Dimension Style dialog box. Choose **Continue** to edit the dimension style as shown in Figure 7-56.

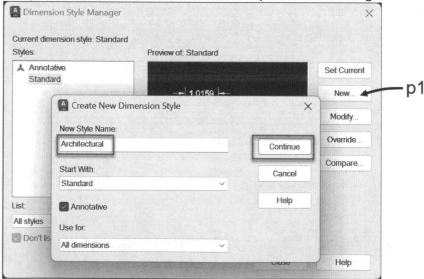

Figure 7-56

14. Click the **Fit** tab. Verify the **Annotative** check box is selected as shown in Figure 7-57.

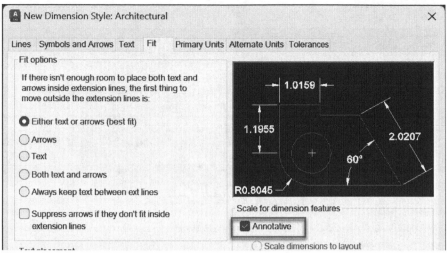

Figure 7-57

15. In this step you will edit the units of the dimension style to Architectural. Select the **Primary Units** tab. Choose **Architectural** from the Unit format flyout and clear the check box for **0** inches in the **Zero suppression** category as shown in Figure 7-58.

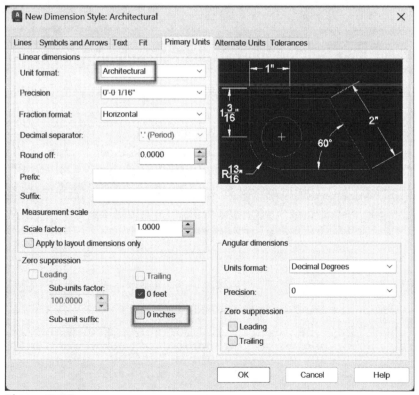

Figure 7-58

16. Choose the **Text** tab. Edit the Vertical text placement to **Above** and choose the **Align with dimension line** radio button in the Text alignment section as shown in Figure 7-59.

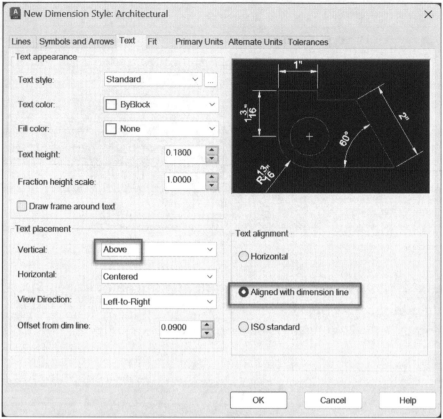

Figure 7-59

17. Click **OK** to dismiss the New Dimension Style dialog box. Choose the Architectural style from the Style list. Choose **Set Current** button as shown in Figure 7-60. The Architectural dimension style is defined as current in the drawing. Choose Close to dismiss the Dimension Style Manager.

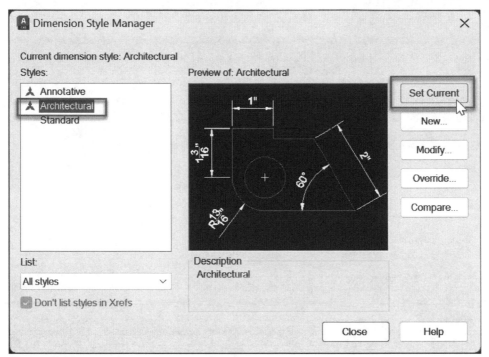

Figure 7-60

18. In the next steps you will determine an appropriate scale for the 36 x 80 rectangle. Choose **Layout 1**. Right click the Layout 1 tab to display the contextual menu as shown in Figure 7-61 and choose **Page Setup Manager**.

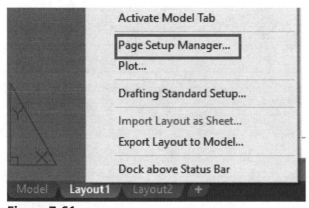

Figure 7-61

19. Choose the **Modify** button to specify a printer for the layout as shown in Figure 7-62.

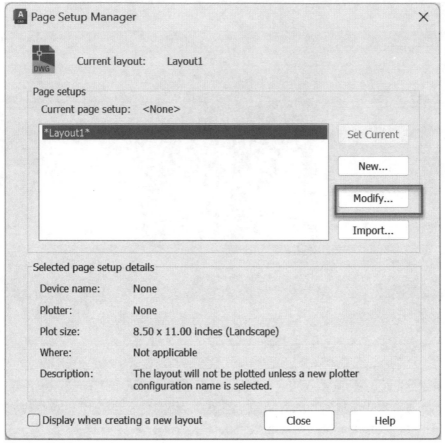

Figure 7-62

20. Edit Printer name to **AutoCAD PDF (General Documentation).pc3** and Paper Size to **ANSI A 11 x 8.5** Inches as shown in Figure 7-63. Choose **OK** to dismiss the Page Setup – Layout 1. Choose **Close** to dismiss the Page Setup Manager.

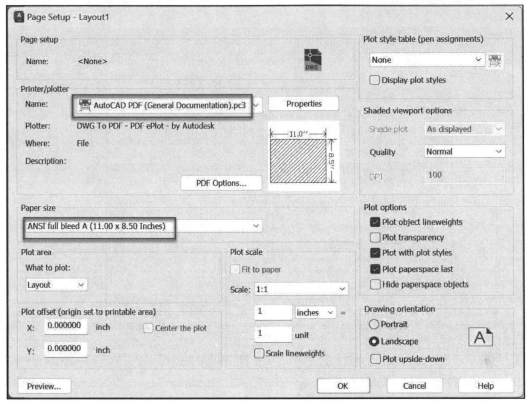

Figure 7-63

21. Double click in the center of the screen to activate Model Space for the viewport as shown in Figure 7-64.

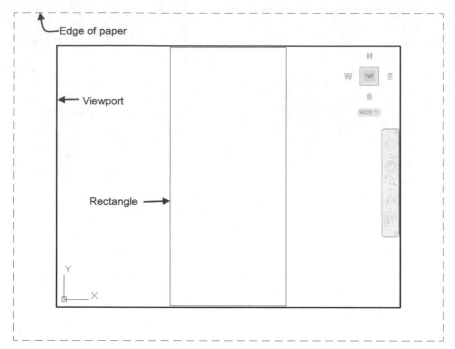

Figure 7-64

22. Choose ½"= **1'-0"** from the Viewport scale as shown in Figure 7-65 of the Status bar.

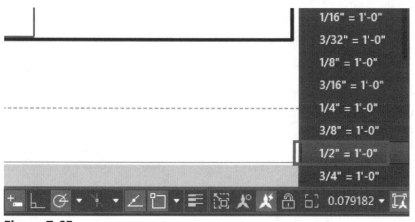

Figure 7-65

23. The revised scaled view of the rectangle is shown in the viewport in Figure 7-66. The ½"=1'-0" scale provides ample space around the rectangle to place dimensions.

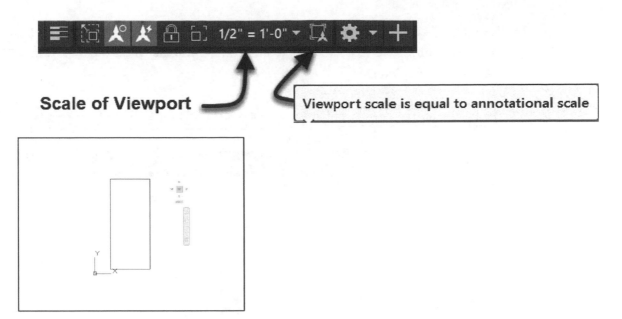

Figure 7-66

24. Click the **Model** tab. Choose the Annotation Scale flyout and choose **½" = 1'-0"** scale as shown in Figure 7-67.

Figure 7-67

25. Choose the **Annotate** tab of the ribbon. Choose **Linear** dimension tool from the Dimension flyout in the Dimensions panel as shown in Figure 7-68.

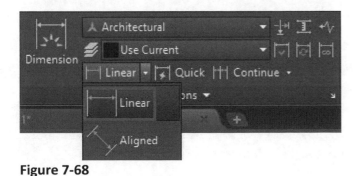

Figure 7-68

26. Note the **½"=1'-0"** scale is shown in the Status bar as shown in Figure 7-69. Verify **Show annotation objects** and **Add scales to annotative objects when the annotation scale changes** are toggled ON.

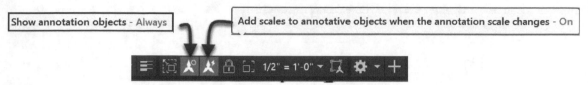

Figure 7-69

27. Add the vertical and horizontal dimensions as shown in Figure 7-70.

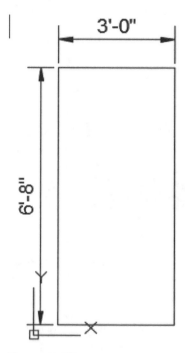

Figure 7-70

28. Choose **Layout 1** to view the rectangle.
29. Note when you Toggle on **Show annotation objects** and **Add scales to annotative objects when the annotation scale changes** in the Status bar as shown in Figure 7-69 the changes to scale representations are applied.
30. Choose the **3/8" = 1'-0"** a scale representation is automatically added to the dimensions as shown in Figure 7-71 since the **Add scales to annotative objects when the annotation scale changes** was toggled ON in the previous step.

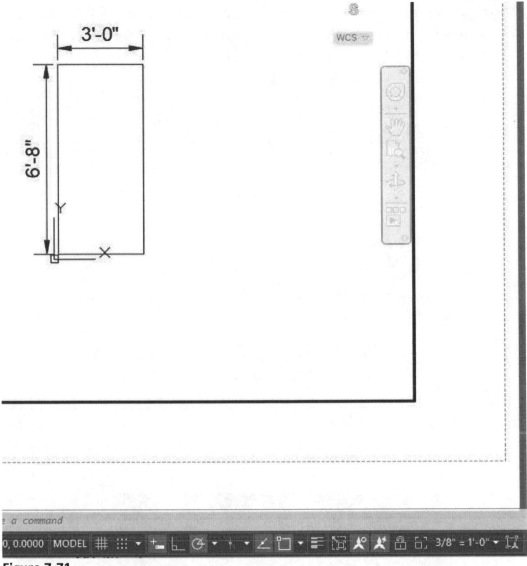

Figure 7-71

31. Select a dimension, right-click and choose **Add/Delete Scales** from the Annotate tab Annotation Scaling panel as shown at p1 in Figure 7-72.

Figure 7-72

32. The scale representations for the dimension are shown in Figure 7-73. The Add / Delete scales allow you to select a scale from the list of scale representation and delete scales if no longer needed. Choose OK to dismiss the dialog box.

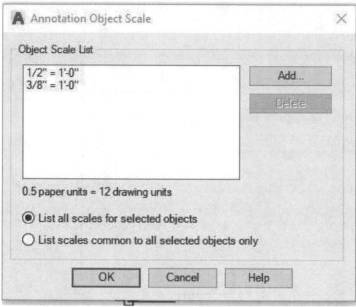

Figure 7-73

33. Note the majority of annotation tools of the Annotate tab consist of styles with the Annotative property. Choose Annotative multi-leader style of the Leaders panel as shown at p1 in Figure 7-74.

Figure 7-74

34. Choose Multileader from the Leaders panel as shown at p2 in Figure 7-74.
    Respond to the workspace prompts as follows:
    Specify leader arrowhead location or [leader Landing first/Content first/Options] <Options>: _nea to (**Select** the rectangle at **p1** in Figure 7-75 to start the leader using the Nearest object snap mode.)
    Specify leader landing location: (**Select** a point near **p2** as shown in Figure 7-75.)

Type **Door Jamb Location** on screen and choose **Close Text Editor** of the Close panel in the Text Editor tab.

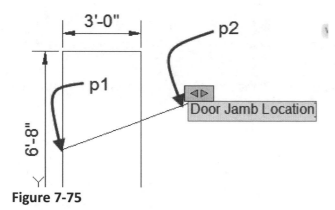

Figure 7-75

35. The multileader is placed as shown in Figure 7-76.

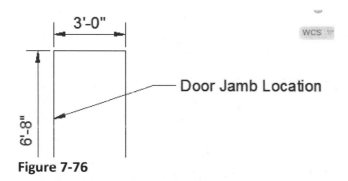

Figure 7-76

36. Choose **Add Leader** of the Leaders panel as shown in Figure 7-77 of Annotate Tab.

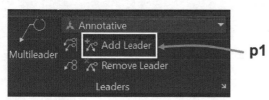

Figure 7-77

Respond to the workspace prompts as follows:

> Select a multileader: (Select the existing multileader at p1 shown in Figure 7-78.)
> 1 found
> Specify leader arrowhead location or [Remove leaders]: _nea to (Hold the **Shift** key, right-click and choose **Nearest** object snap mode. Select a location at p2 as shown in Figure 7-78.)
> Specify leader arrowhead location or [Remove leaders]: (Press **Enter** to end the command.)

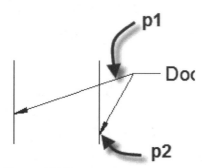

Figure 7-78

37. Double click on the **Door Jamb Location** text and add an **s** to the Location word as shown in Figure 7-79. Choose Close Text Editor of the Close panel.

Figure 7-79

38. The completed drawing is shown in Figure 7-80. Choose Save in the Quick Access toolbar. Navigate to your student directory and type **Annotative Dims and Leaders Your Name** as the drawing name.

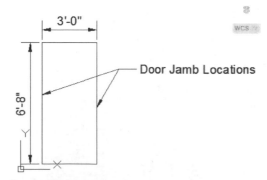

Figure 7-80

7.4 Multi-Line Text Justification Test   Name_____

The following questions will test your ability to identify the mtext justification based upon the grip display when editing justification in the Properties dialog box.

1. Circle the text justification setting for the grip display of the text shown at left.

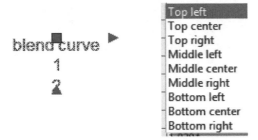

2. Circle the text justification setting for the grip display of the text shown at left.

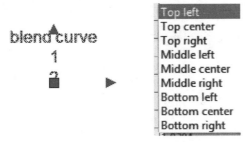

3. Circle the text justification setting for the grip display of the text shown at left.

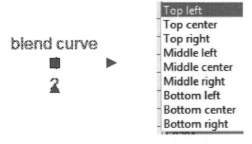

4. Circle the text justification setting for the grip display of the text shown at left.

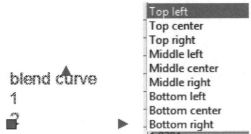

5. Circle the text justification setting for the grip display of the text shown at left.

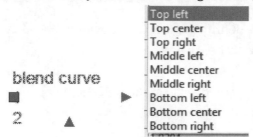

6. Circle the text justification setting for the grip display of the text shown at left.

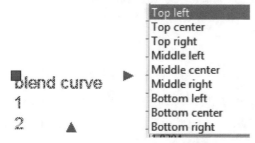

7. Circle the text justification setting for the grip display of the text shown at left.

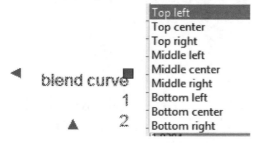

8. Circle the text justification setting for the grip display of the text shown at left.

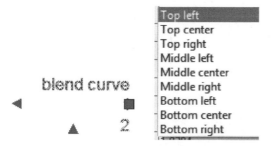

9. Circle the text justification setting for the grip display of the text shown at left.

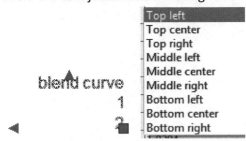

10. Circle the text justification setting for the grip display of text shown at left.

.Text

Bottom Left

Center

Bottom Right

Middle

Notes:

Chapter 8
Reusing Existing Content

This chapter will review the methods of creating and inserting blocks. The editing of existing block definitions using the Block Editor is included. The use of the DesignCenter to import blocks from other drawing and creating tool palettes will be introduced.

The student will be able to:

1. Create block content with specific basepoint location and control of color and linetype per layer
2. Insert blocks from the Block palette, block gallery or using the Insert dialog box
3. Use the Block Editor to change graphics, color, layer and linetype of blocks
4. Place blocks of a drawing on tool palettes for use in other drawings
5. Insert blocks from the DesignCenter and create tool palettes from blocks of a drawing in the DesignCenter
6. Use the Count command to display the number of blocks or entities of a drawing in a table

Creating and Inserting Blocks

Blocks provide the tools to quickly place multiple entities into a drawing as one object. Blocks reduce the drawing size since each insertion references the block definition. If a block consists of 10 lines each additional insertion only adds 1 unit to the drawing size. The content of a block is typically symbols, furniture or any graphics used repetitively in a drawing.

Controlling Block Properties

If your settings for color, linetype and lineweight are set to bylayer and need all blocks inserted on a layer to display with the color, linetype and lineweight of the layer you should create blocks from content placed on layer 0. If block content is created on Layer 0 and the block is inserted in a drawing with bylayer properties, the block will assume the color and linetype properties of the layer that it is inserted on. Blocks will retain their unique color, linetype or lineweight if not created on layer 0 and with properties not set to bylayer or byblock. You can use the Block Editor to edit layer, linetype, color and lineweight of block components.

The Block commands are located on the Home tab in the Block panel and on the Insert tab in the Block and Block Definition panels as shown in Figure 8.1. To create a block choose the Create Block command as shown at p1 in Figure 8.1.

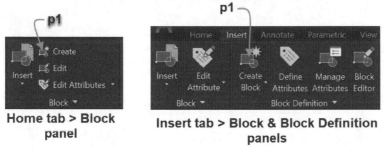

Home tab > Block
panel

Insert tab > Block & Block Definition
panels

Figure 8-1

When you choose the Block command, the Block Definition dialog box opens which allows you to insert the unique name for the block as shown at p1 in Figure 8-2. The base point selection is chosen using the Pick point button shown at p2 in Figure 8-2. Choosing this button returns you to the graphic screen which allows you to use object snap tools to select the base point. The base point location is used as the handle when inserting blocks. Note the Annotative check box allows you to create blocks that are sized by the annotative scale factor. If the Convert to block radio button is ON in the Objects section, the entities used to create the block are converted to a block when you click OK.

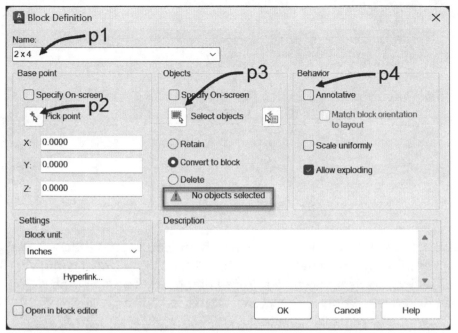

Figure 8-2

The technique to create and insert a block is described below; you can open the **Block demo** file from the **AutoCAD 2024 Certified User Exercise Files \ Exercise Files \ Certification Demo Files \ Ch 8** to try this technique as you read the steps.

1. The content for creating a block for detailing to represent the section of view of a 2 x 4 framing member is shown at left in Figure 8-3.
2. The content of the block was created on layer 0 as shown in Figure 8-3.
3. Choose the Block command from the Block panel in the Home tab to open the Block Definition dialog box. Type **2 x 4** in the Name field shown at p1 to specify the name. In the Base point section choose the **Pick Point** button as shown at p2 in Figure 8-3. The display of the Block Definition dialog box is closed to allow you to choose the endpoint of the lines shown at p3 to specify the Base point of the new block. When you select the location the Block Definition dialog box reopens.

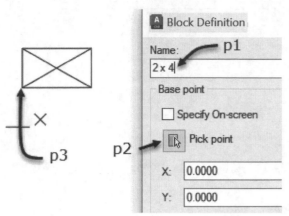

Figure 8-3

4. Choose the **Select objects** button shown at p1 in the Objects section of the Block Definition dialog box of Figure 8-4. Note, the Block Definition dialog box temporarily closes. Click at p2 and p3 to select the entities. Press Enter to end selection—the Block Definition dialog box reopens. Verify 3 objects are selected as shown in Figure 8-4. Choose OK to dismiss the Block Definition dialog box.

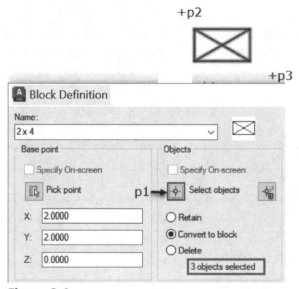

Figure 8-4

5. Verify Layer 0 is current. To insert an instance of the block choose Insert shown at p1 in Figure 8-5 from the Block panel of the Home tab. The Block Gallery shown in Figure 8-5 opens which includes blocks of the drawing. The blocks of a drawing are displayed in the Block Gallery space when they are created. When you click on the 2 x 4 block shown at p2 in Figure 8-5, an instance of the block is locked to the cursor; as shown at p3 you can click to insert the block in the drawing. You can press the Down Arrow key at p4 as shown in Figure 8-5 to change the options prior to inserting the block. The Basepoint option allows you to preset a different basepoint during the insertion of a block.

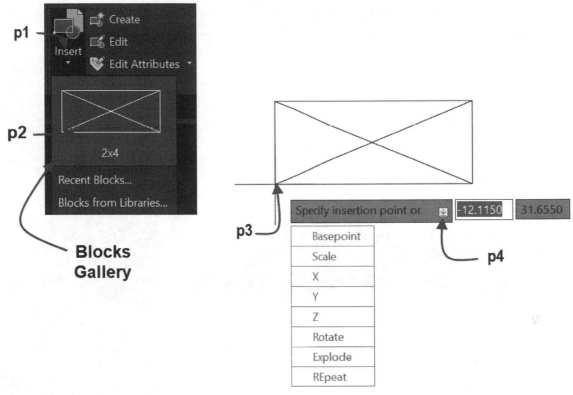

Figure 8-5

6. Note in the step you inserted the block on layer 0. Change the layer to Details which has cyan color and insert the block. The block should appear with cyan color.

Blocks Palette

The Blocks Palette provides four tabs of blocks used in the current drawing, recent drawings, favorites, and a tab for a library of blocks. To open the Blocks palette, choose Blocks from the Palettes panel in the View tab of the ribbon as shown at p1 in Figure 8-6. The Current Drawing tab allows you to view the blocks that are listed in the Blocks Gallery of the current drawing. You can insert blocks into a drawing from the Blocks Gallery or the Blocks palette. In the Blocks palette you can click a block listed in the Current Drawing tab to insert the block. If you right click a block, you can choose either to Insert or Insert and Explode options. Note, if a block is exploded it is no longer a block. Insertion Options presets are included on each tab and should be set prior to selecting a block for insertion. If you preset the rotation angle and scale values to insert, all future blocks are inserted using the same values. The Insertion Options check boxes include **Repeat Placement** as shown at p2 in Figure 8-6 which allows you to add multiple insertions after the first insertion without requiring you to reselect the block.

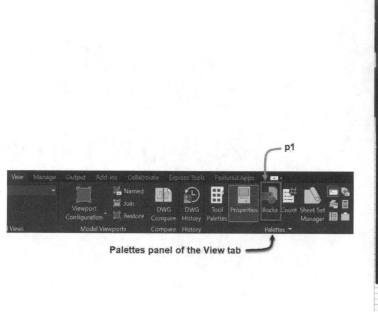
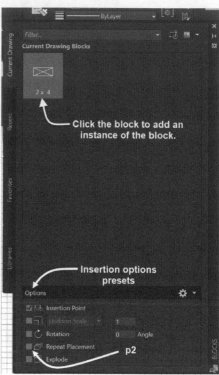

Figure 8-6

Note, if you choose the Recent tab as shown in Figure 8-7, this tab shows blocks from the current drawing and blocks used in other drawing files during the session. You add blocks to your Favorites by selecting a block from the recent tab and choose Add to Favorites as shown at p1 in Figure 8-7.

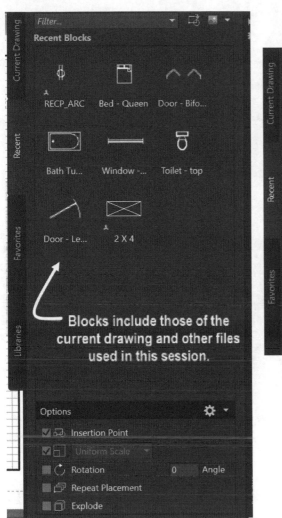

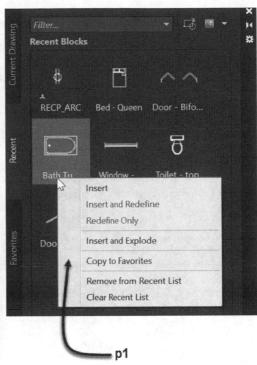

Figure 8-7

All four tabs of the Blocks palette allow you to control the format for the block list as shown at p1 in Figure 8-8. The Libraries tab allows you to access other closed drawings using the Browse button shown at p2 in Figure 8-8. If you choose the Blocks from Libraries tab as shown in Figure 8-8, you can access the blocks from other files on your computer or from network in a dialog box. The dialog box shown in Figure 8-8 shows the path to access the House Designer.dwg which provides sample blocks for a house included with AutoCAD software.

C:\Program Files\Autodesk\AutoCAD 2022\Sample\en-us\DesignCenter\House Designer.dwg\Blocks (20 Item(s))

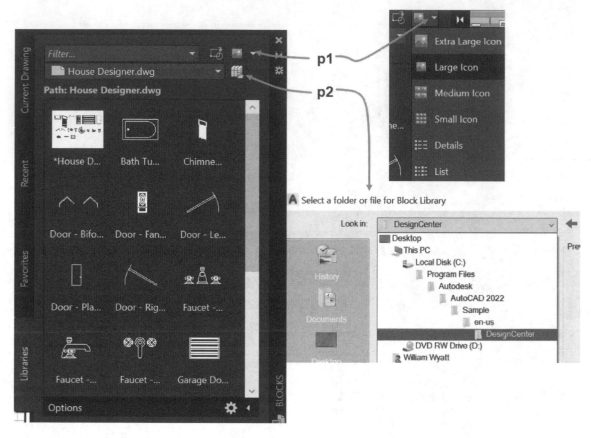

Figure 8-8

Editing Blocks

The blocks in a drawing can be modified using the Block Editor. If you select an instance of a block and choose the Block Editor as shown in Figure 8-9, the Edit Block Definition opens as shown at right in Figure 8-9. After specifying the block click OK to dismiss the Edit Block Definition dialog box.

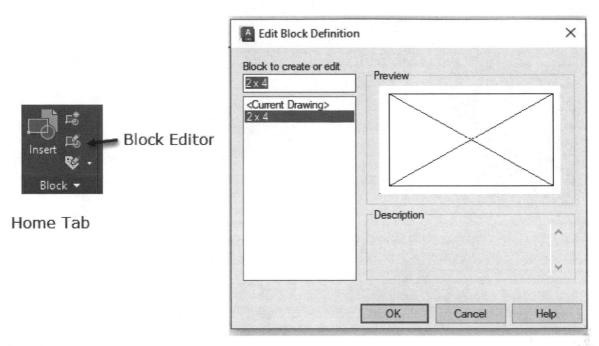

Figure 8-9

The Block Editor tab opens as shown in Figure 8-10. You can open the Home tab and add entities from the Draw panel or edit entities using the Modify panel. Upon completion of the edit you can choose Save Block from the Open Save panel and the changes will apply to all blocks of the same name in the drawing when you choose Close Block Editor.

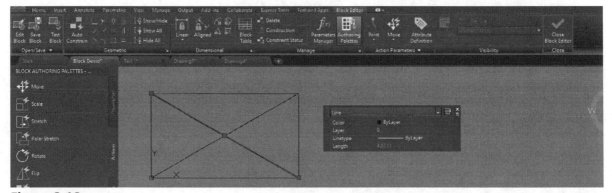

Figure 8-10

The Block Editor can be used to create a new block if upon completion of the edit you choose Save Block As from the extended Open Save panel shown at p1 in Figure 8-11. When you enter a unique name, click OK to dismiss the Save Block As dialog box and choose **Close Block Editor** from the Block Editor tab.

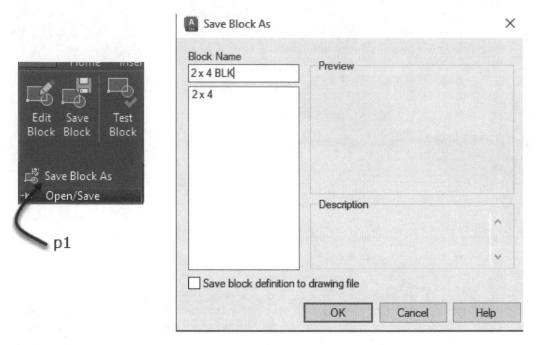

Figure 8-11

In addition to adding or deleting entities from a block definition in the Block Editor, you can verify or change the layer of entities of a block to Layer 0 and the properties to bylayer of all components of a block using the Properties palette as shown in Figure 8-12.

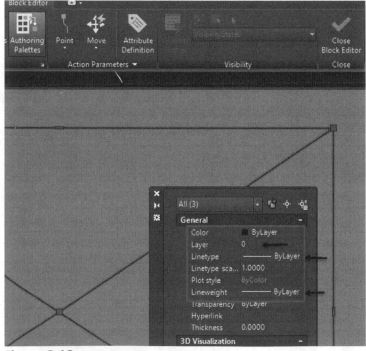

Figure 8-12

Using the DesignCenter

The DesignCenter allows you to access Blocks from other drawings and create tool palettes of blocks in the drawings. Choose the DesignCenter from the Palettes panel in the View tab as shown at p1 in Figure 8-13.

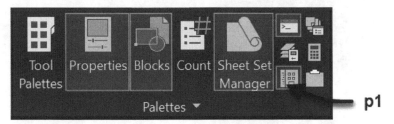

Figure 8-13

When Design Center opens as shown in Figure 8-14, there are three tabs as shown at p1. The Open Drawings tab allows you to view and insert content from any open drawing. The 2 x 4 block shown at p2 can be dragged to any open drawings from the Block Demo drawing as shown in Figure 8-14.

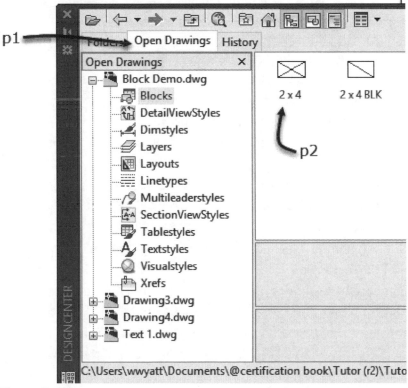

Figure 8-14

To create a tool palette of the blocks in a drawing choose Blocks category of a drawing as shown at p1, right click and choose Create Tool Palette as shown in Figure 8-15. The tool palette of the Block Demo file is shown at right in Figure 8-15. Blocks of a drawing can be selected and dragged to any tool palette that is displayed in AutoCAD.

Figure 8-15

8.1 Creating Blocks and Tool Palettes

The following tutorial provides an exercise in the creation of blocks and the use of the Insert command and tool palettes.

1. Open AutoCAD 2024 from the Desktop shortcut.
2. Choose New from the Quick Access toolbar. Verify acad.dwt is the template; choose Open to dismiss the Select Template dialog box.
3. Save the drawing in your student folder. Name the file **Blocks.dwg**
4. Toggle ON Object Snaps, Dynamic Input, Polar Tracking, and Object Snap Tracking in the Status bar.
5. Choose the **Rectang** command from the Draw panel of the Home tab.
   Respond to the workspace prompts as follows:
   Command: _rectang
   Specify first corner point or [Chamfer/Elevation/Fillet/Thickness/Width]: **2,2** (Type 2,2 to specify the first corner of the rectangle.)
   Specify other corner point or [Area/Dimensions/Rotation]: **d** (Press the Down Arrow key on the keyboard to toggle to Dimensions. Press Enter to select the dimensions option.)
   Specify length for rectangles <10.0000>: **3** (Type 3 to enter the length of the rectangle.)
   Specify width for rectangles <10.0000>: **2** (Type 2 to enter the width of the rectangle.)
   Specify other corner point or [Area/Dimensions/Rotation]: (Left click a location to the right and above the rectangle as shown in Figure 8-16.)

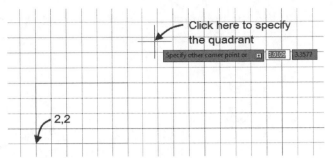

Figure 8-16

6. Choose the Line command from the Draw panel of the Home tab and draw lines from each of the corners of the rectangle as shown in Figure 8-17.

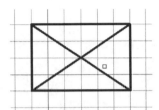

Figure 8-17

7. Choose the Home tab. Choose the **Create Block** tool from the Block panel as shown in Figure 8-18 to open the Block Definition dialog box.

Home Tab

Figure 8-18

8. Edit the Block Definition as follows and type **Chase** in the Name field. Choose the **Pick point** button as shown in Figure 8-19 to specify the base point for the block.

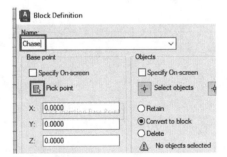

Figure 8-19

9. When the graphics window opens select the lower left corner of the rectangle as shown at **p1** in Figure 8-20.

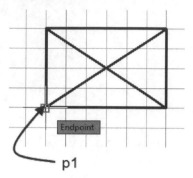

Figure 8-20

10. When the Block Definition dialog box reopens choose the **Select objects** button as shown in Figure 8-21.

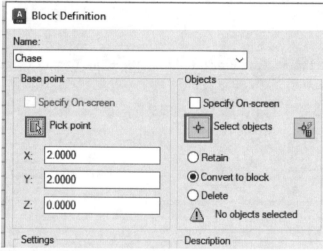

Figure 8-21

11. When the graphic window reopens left click at p1 and p2 as shown in Figure 8-22 to create a crossing selection window. Press Enter to end the selection.

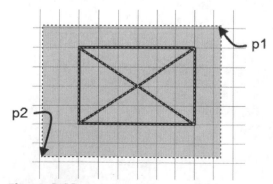

Figure 8-22

12. The Block Definition dialog box reopens; verify the name, insertion base point and Objects and Behavior are edited as shown in Figure 8-23. Choose OK to dismiss the dialog box.

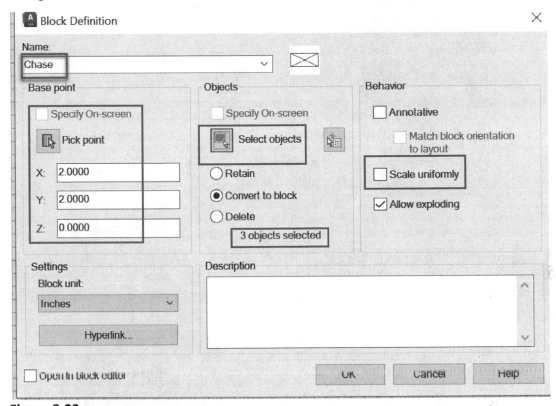

Figure 8-23

13. Choose Layer Properties from the Layers panel of the Home tab as shown in Figure 8-24.

Figure 8-24

14. Create a new layer named Layer 1 as shown in Figure 8-25. Edit the color to Blue. Set Layer 1 current.

Figure 8-25

15. Expand the Insert command of the Block panel in the Home tab. Choose Chase from the block gallery as shown in Figure 8-26.

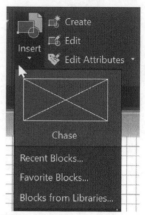

Figure 8-26

Move to the workspace and left click to place an instance of the block as shown at left in Figure 8-27. To add additional instances of a block, choose the chevron shown at p1 to open the Options of the Blocks palette. Choose the **Repeat Placement** as shown at p2 and choose the Chase to add two additional blocks to the drawing as shown in Figure 8-27.

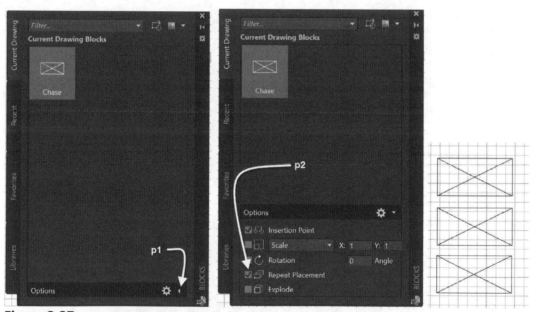

Figure 8-27

16. Edit the Y Scale value to .5 to preset the scale in the Options section of the Current Drawing tab as shown at p1 in Figure 8-28.

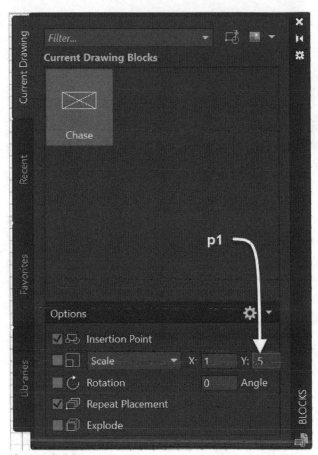

Figure 8-28

17. Move the cursor to the workspace and insert an instance of the block as shown at p1 and p2 in Figure 8-29.

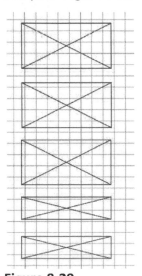

Figure 8-29

18. Choose the **View** tab. Choose the **Tool Palettes** tool of the Palettes panel as shown at p1 in Figure 8-30.

Figure 8-30

19. Choose the Architectural palette as shown in Figure 8-31. This palette consists of existing blocks that are provided with the software.

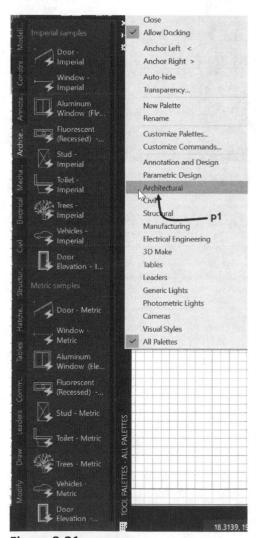

Figure 8-31

20. Choose **Save** from the Quick Access toolbar. Save the drawing as Block Your Name in your student folder. Saving the drawing at this point is important since in the next steps you will place a tool on the palette that references the block definition of this drawing.

21. In this step you will add the Chase block to the palette. Select the Chase block to display its grip. Move the cursor over an entity of the block, then left click and drag the cursor to the palette as shown in Figure 8-32. (Note, do not select the grip of the block.)

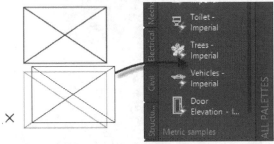

Figure 8-32

22. The image of the Chase block will appear in the palette as shown in Figure 8-33.

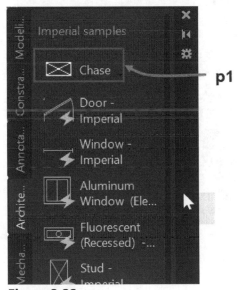

Figure 8-33

23. The Chase block may be used for future drawings if the Block.dwg file remains on your computer at the location where you saved it to support the block.

24. Note, you may draw lines on specified layers and drag the line to a tool palette to create a tool for the Line command with the same linetype and layer. Create a new layer named Center. Set the Center layer linetype to center and color to green. Set the Center layer current as shown in Figure 8-34.

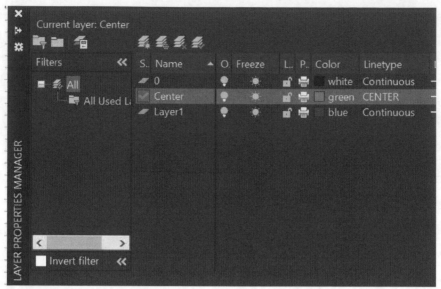

Figure 8-34

25. Draw a line on the Center layer. Drag the center line onto the tool palette. Right-click the new tool and choose Rename as shown at p1, type Center to name the tool. The Center line tool is shown at p2 in Figure 8-35.

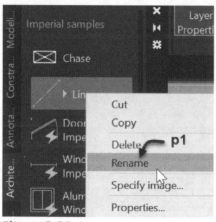 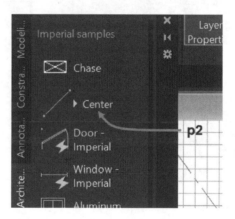

Figure 8-35

26. Additional blocks may be accessed from the DesignCenter. You may choose the DesignCenter from the Palettes panel in the View tab as shown in Figure 8-36 to open the DesignCenter.

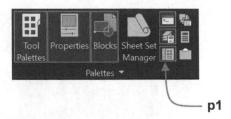

p1

Figure 8-36

27. The content of the DesignCenter includes various drawings with blocks as shown in Figure 8-37. The path to the Home of the DesignCenter is shown at p1 and you can press the Home button as shown in Figure 8-37 to access the path.

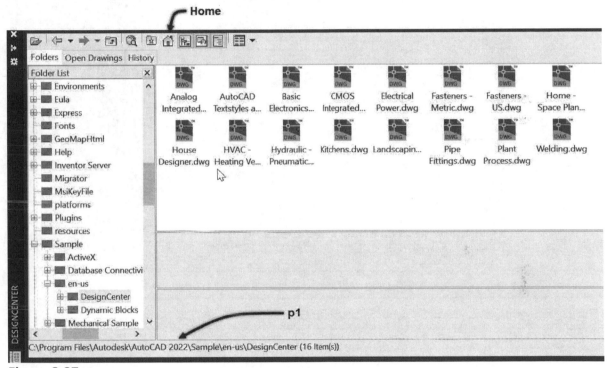

Figure 8-37

28. Note, the DesignCenter is located at C:\Program Files\Autodesk\AutoCAD 2024\Sample\en-us\DesignCenter in AutoCAD 2024 on your computer.

29. You may choose the Open Drawings tab to select your current drawing file within the DesignCenter; right-click and choose Create Tool Palette as shown in Figure 8-38 to generate a palette of all blocks within the drawing.

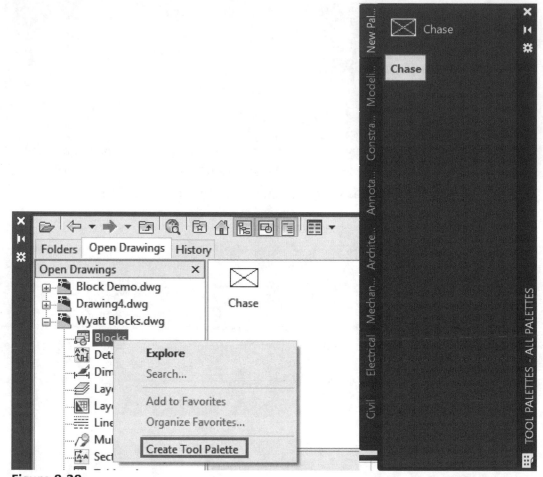

Figure 8-38

30. Save the drawing as Blocks_Your Name in your student folder. Close the drawing file.

Tutorial 8-2 Editing a Block Assignment

This tutorial assignment requires the use of the Block Editor to change the content and color property of a block.

1. Open the Block Edit file of the AutoCAD 2024 Certified User Exercise Files \ Exercise Files \ Ch 8 folder.
2. The drawing consists of a block named Recpt. Insert the block on the Electrical layer. Edit the block by deleting the line shown in Figure 8-39.

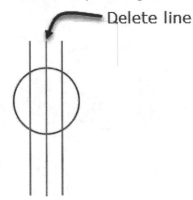

Delete line

Figure 8-39

3. Edit the color of the block such that when inserted on the Electrical layer the color of the block will be blue.
4. Insert five instances of the Recpt block.
5. Save the file as Your Name Block Edit and close the drawing.

Counting blocks and entities

AutoCAD 2024 includes the new Count command which can be used to count similar entities or blocks of a drawing. The results of the count can be placed as a text field in a note or in an AutoCAD table to create a schedule. You can access the Count command from the shortcut menu without selecting any objects as shown at p1 in Figure 8-40 or from the shortcut menu with one or more objects selected as shown at p2 in Figure 8-40.

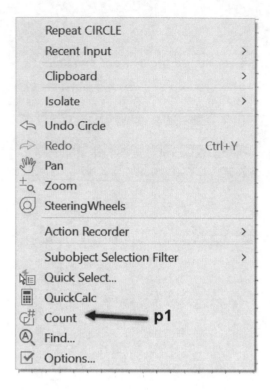

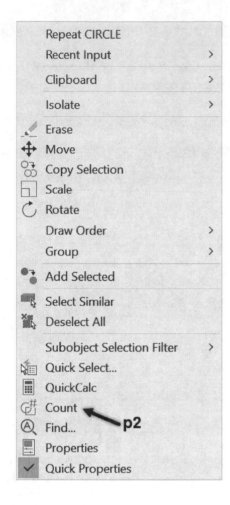

Figure 8-40

When you select an entity or a block the count toolbar is displayed at the top center in the workspace which lists the total number of objects that exist in the drawing as shown at p1 in Figure 8-41. Following quantity value is a blue circle with the letter "i" value indicating there are no errors such as two entities inserted in same place. The arrows shown at p2 in Figure 8-41 will zoom the graphics window to each instance and the instance is shown in green highlight. To insert the text field of the quantity choose the button shown at p3 in Figure 8-41. To exit the Count command, choose the green check button shown at p4 in Figure 8-41.

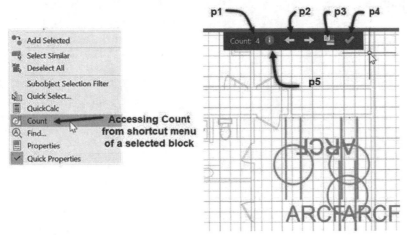

Figure 8-41

If multiple blocks or entities exist in the same place, the duplication will be highlighted in red as shown at p1 in Figure 8-42, and the triangle warning symbol shown at p2 in Figure 8-42 replaces the blue information button as shown at p5 in Figure 8-41.

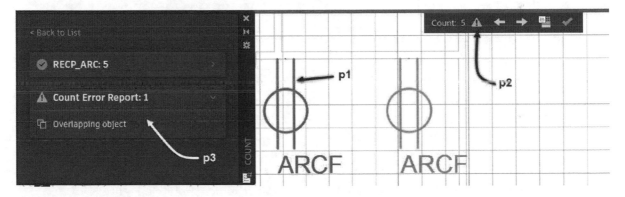

Figure 8-42

Tutorial 8-3 Using the Count tool to create a schedule.

The following tutorial provides an exercise in the creation of blocks and the use of the Count command and the creation of an annotative block. The results of the count command will be displayed as a text field and a table.

1. Verify Drafting and Annotation is the current workspace as shown in the Quick Access toolbar.
2. Choose Open from the Quick Access toolbar.
3. Navigate to the AutoCAD 2024 Certified User Exercise / Exercise Files / Ch 8 folder.

4. Select the Count Floorplan.dwg file. The file consists of lines which represent the walls of the building and a Floor Plan and Block Contents Layouts. The floorplan layout is locked and scaled at 1/8" = 1' - 1'-0" as shown in Figure 8-43.

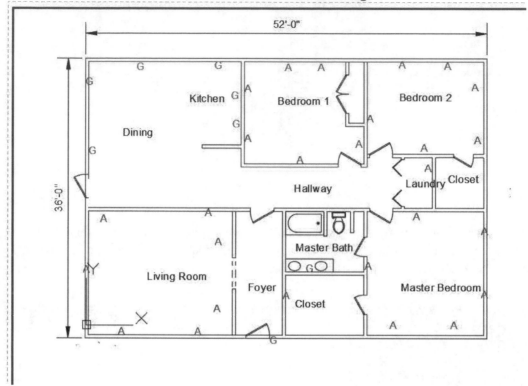

Figure 8-43

5. Open the Block Contents layout which is scaled to 1:1 for the creation of annotative blocks. The graphics shown have been created actual size of a symbol. The application of annotation scaling will be applied for typical architectural symbols in model space.

6. Create the AutoCAD block named AFCI using the graphics shown in Figure 8-44.

7. Choose the Create Block tool from the Home tab, Block panel as shown in Figure 8-44. Type the name AFCI in the Name field shown at p1 in Figure 8-44.

8. To specify the Basepoint choose the Pick point button shown at p2 in Figure 8-44. When the graphics window opens choose the AutoCAD point using the node object snap mode shown at p3 in Figure 8-44.

9. To specify the graphics for the block choose the Select objects button shown at p4 in Figure 8-44. When the graphics window opens select the graphics using a crossing selection from p5 to p6 in Figure 8-44. Check the Retain radio button and clear Delete and Convert to block check boxes as shown at p7 in Figure 8-44.

10. Check the Annotative radio button shown at p8 in the Behavior section of Figure 8-44.

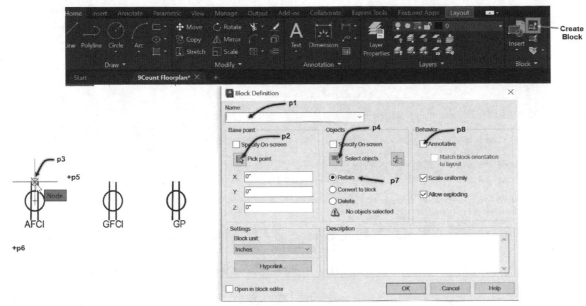

Figure 8-44

11. Verify the settings as shown in Figure 8-45 then choose the OK button.

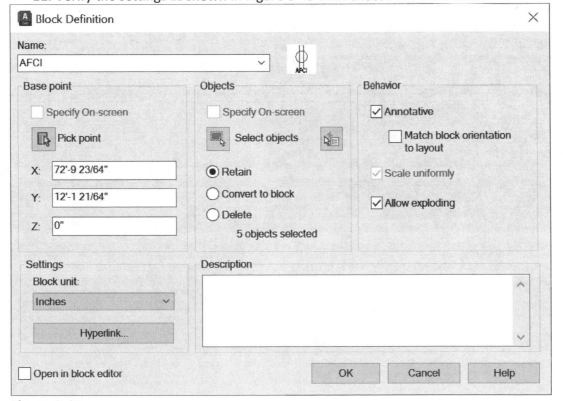

Figure 8-45

12. Choose the Floor Plan layout.

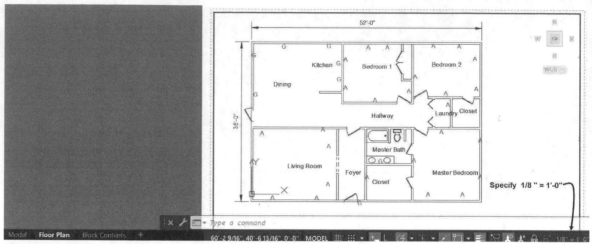

Figure 8-46

13. Verify model space is current and the scale of the viewport is set to 1/8" = 1'– 0" as shown in Figure 8-46.
14. In the next series of steps you will insert the AFCI (Arc Fault Circuit Interrupter) receptacle block in Bedroom 1.
15. Toggle on the Nearest object snap mode from the Object Snap flyout.
16. Set the electrical layer current.
17. Choose the Insert command from the Block panel. Select the AFCI block as shown in the Blocks Browser of Figure 8-47.

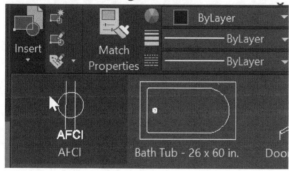

Figure 8-47

18. Set the rotation as necessary to insert receptacles as shown in Figure 8-48 in Bedroom 1.

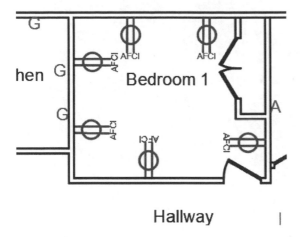

Figure 8-48

19. Choose a receptacle in Bedroom 1, right click and choose Select Similar to create a selection set of the receptacles as shown in Figure 8-49.

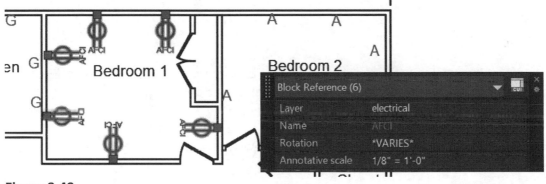

Figure 8-49

20. Right click and choose the Count command from the shortcut menu as shown at p1 in Figure 8-50. Choose the Insert Field command from the Count toolbar as shown at p2 in Figure 8-50.

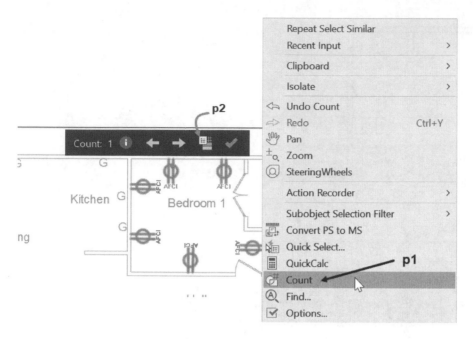

Figure 8-50

21. The default settings for the field text is not set to annotative text; therefore, select a point below the Bedroom 1 room title to display the grip as shown in Figure 8-51, right click the grip and edit the Mtext settings as shown in Figure 8-51 at p1 to Annotative to YES, Annotative scale to 1/8" = 1'-0" as shown at p2, and Paper text height to 1/8" as shown at p3. Choose the green check mark shown at p4 in Figure 8-51 to display the count.

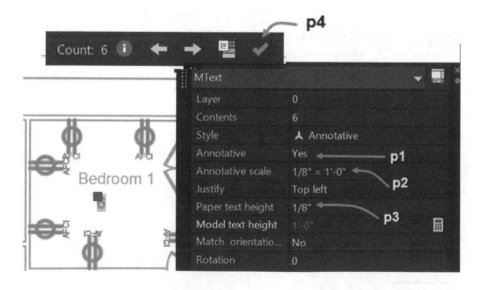

Figure 8-51

22. Choose Count tool from the Palettes panel to open the Count palette. Choose the button shown at p1 in Figure 8-52 to create a table that includes the list of blocks and the quantities included in the count. Choose the box shown at p2 to specify the AFCI blocks.

23. Choose the Insert button shown at p3 in Figure 8-52.

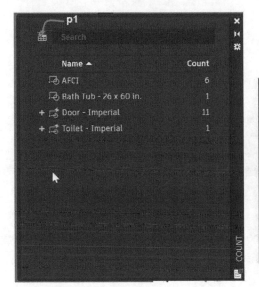

Figure 8-52

24. Select a location below the floorplan as shown at p1 in Figure 8-53.

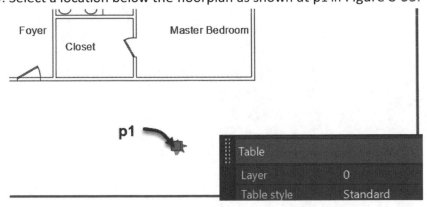

Figure 8-53

25. Retain the selection of the location shown at p1 in Figure 8-53. Choose the Scale command from the Modify panel. Respond to command prompts as follows.

> Command: _scale 1 found
>
> Specify base point: Choose the point shown at p1 in Figure 8-53.

Specify scale factor or [Copy/Reference]: **96** Enter (Type 96 and press Enter to scale the table to the 1/8" = 1'-0" scale). The table and the count values are shown in Figure 8-54.

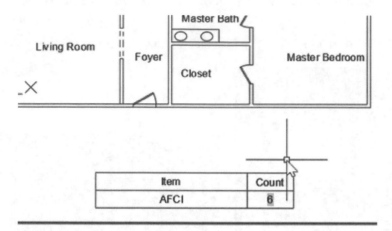

Figure 8-54

Note: If you add or delete AFCI receptacles in the drawing the table and the text field will update. Type REGEN in the command line to be certain the count is the latest for the drawing.

Chapter 9

Layouts and Printing

This chapter will review the methods of printing and the use of Viewports and Layouts. The use of Page Setups for creating various paper sizes and scaling of content is introduced.

The student will be able to:

1. Print to scale from Model or a Layout
2. Print the content to a file
3. Create rectangular and irregular shaped viewports
4. Create Page Setups for printing layouts
5. Control annotation display per viewport
6. Scale the viewport content

Printing

Printing to paper or an Adobe file is often used to distribute the content of designs within the engineering community. Printing a scaled drawing of content from the Model tab is accomplished by specifying a scale factor in the Plot-Model dialog box. Printing content from a Layout tab the print scale factor is 1:1, and the image shown in a viewport is scaled using the Viewport Scale factor of the Status bar. To print a design from the Model choose Print from the Quick Access toolbar and open the Plot dialog box as shown in Figure 9-1.

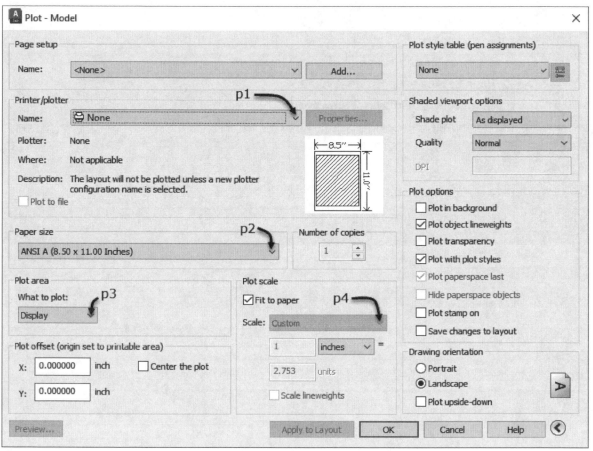

Figure 9-1

Edit the following sections of the Plot-Model dialog box:

Printer/plotter-Choose from flyout list of *Name* field shown at p1 in Figure 9-1 that includes all hardware installed on the computer.

Paper size-The *Paper size* flyout shown at p2 in Figure 9-1 includes paper sizes supported as per the name of printer /plotter selected.

Plot area- The *What to plot* flyout options shown at p3 in Figure 9-1 are Display, Extents, Limits and Window. The display option will include entities of workspace as shown when you start the print command. The extents option includes all entities regardless of current visibility when printing and limits option prints the entities which reside within the limit settings. The Window option when selected temporarily closes the Plot dialog box and prompts you to specify the corners of a rectangular space in the workspace for printing. Note, when plotting from a layout the Plot Layout dialog includes a default Layout option which prints all content of the layout.

Plot scale- Includes a Fit to paper check box which turns off scaling when checked. If the Fit to paper is not checked, you can select the scale factor from the scale flyout shown at p4 in Figure 9-1.

Preview – When you select the Preview button the content of the final print will be shown in the workspace.

Viewports

Model viewports can be created in the Model tab to assist in the visualization of three-dimensional parts. To create viewports in the Model tab, choose the Viewport Configuration flyout shown at p1 in Figure 9-2 from the Model Viewports panel in the View tab.

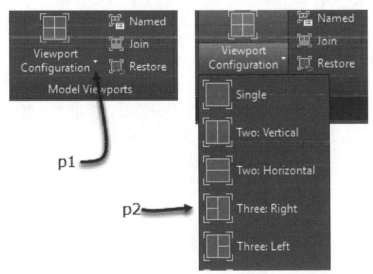

Figure 9-2

If you activate a Layout tab using the acad.dwt template a viewport is included; however, additional viewports may be needed for printing. When you activate a Layout, a Layout tab is added to the ribbon as shown in Figure 9-3. The tab includes the Layout Viewports panel as shown at p1 and the Viewports command flyout shown at p2 in Figure 9-3. Selecting the Viewports command flyout allows you to create three different viewport configurations as shown in Figure 9-4. The launching arrow shown at p3 as shown in Figure 9-3 opens the Viewports dialog which can be used to create predefined viewport configuration.

Figure 9-3

The Rectangular option allows you to create the viewport by selecting two corners of the rectangle. The Polygonal option allows you to create an irregular shape using line and arc segments. You can create a polygon, circle, ellipse or closed polyline and convert the closed shape to a viewport using the Object option of the Vports command.

Figure 9-4

The launching arrow shown at p3 in Figure 9-3 opens the Viewports dialog box which can be used to create viewports based upon predefined configurations for 2D and 3D designs as shown in Figure 9-5.

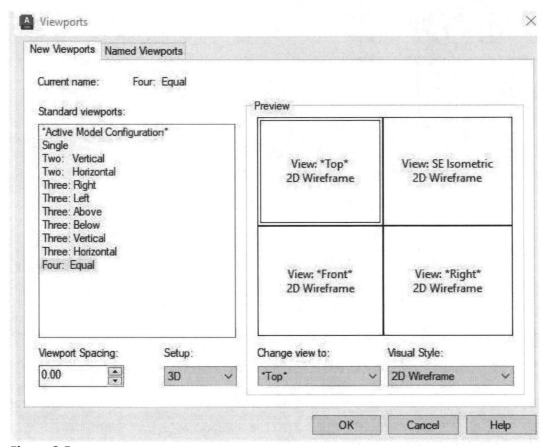

Figure 9-5

Printing from Layouts

The viewport is used as a window to look at a design part at a scale factor. The size of a layout is dependent upon available printer or plotter capabilities. The size of the layout is specified using the Page Setup Manager. The Page Setup Manager allows you to print to paper or to a file to fit the design part to an appropriate scale. The layouts of the acad.dwt template are for A size paper; therefore, the page size should be specified before setting viewport size and shape using the Page Setup Manager.

To open the Page Setup Manager, choose Page Setup of the Layout panel in the Layout tab as shown at p1 in Figure 9-6. You can also right-click the Layout tab and choose Page Setup as shown at p2 in Figure 9-6 to open the Page Setup Manager.

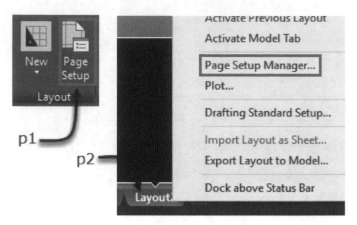

Figure 9-6

When the Page Setup Manager opens choose the New button shown at p1 in Figure 9-7. Enter the name for the new Page Setup as shown at p2 in Figure 9-7. Note you can choose the Import button as shown in Figure 9-7 and import page setups from other drawings.

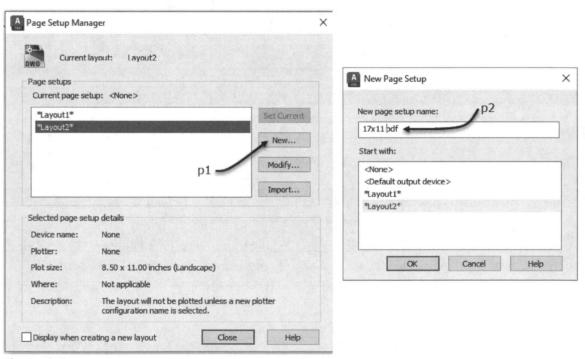

Figure 9-7

Set the printer by choosing from the flyout list as shown at p1 in Figure 9-8. The flyout list is dependent upon printers or plotter installed on the computer. Choose from the flyout at p2 in Figure 9-8 to specify the paper size from the list. When plotting from a layout verify the plot area is set to Layout as shown at p3 in Figure 9-8 and the plot scale is set to 1:1, shown at p4 in Figure 9-8 since the Viewport Scale factor will control the scaling. Choose the radio button for

Landscape in drawing orientation section shown at p5 in Figure 9-8. Finally, choose the Preview button shown at p6 in Figure 9-8 to view the projected print. Choose OK to dismiss the dialog box and return to the Page Setup Manager as shown in Figure 9-9. Choose the 17 x 11 pdf page setup and choose the Set Current button as shown in Figure 9-9. Choose Close to dismiss the Page Setup Manager. The layout has changed to 17 x 11 as shown in Figure 9-10.

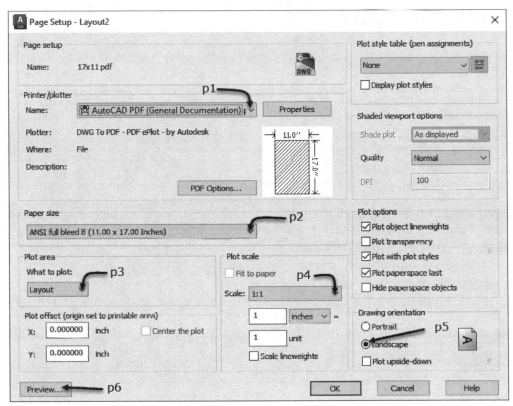

Figure 9-8

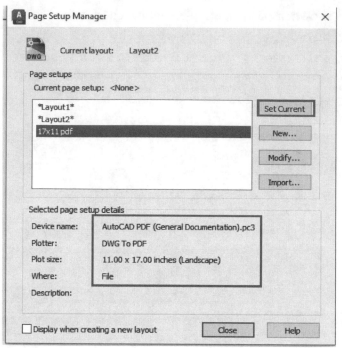

Figure 9-9

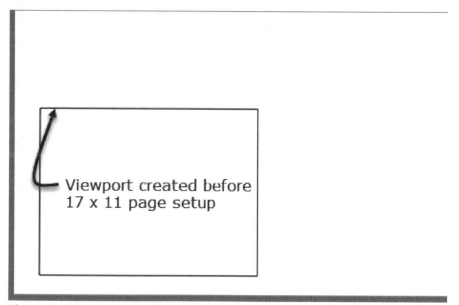

Viewport created before
17 x 11 page setup

Figure 9-10

Creating Layouts and Viewports

The following tutorial includes content regarding the edit of a layout, creating a polygonal viewport and setting the scale per viewport.

1. Click New > AutoCAD DWG from the Application Menu. Specify ACAD.dwt as the template for the new drawing.
2. Verify the Drafting & Annotation workspace is current in the Quick Access toolbar.
3. Choose the Home tab. Choose Layer Properties from the Layers panel of the Home tab as shown in Figure 9-11.

Figure 9-11

4. Create the following layers with associated properties:

Name	Color	Linetype	Plot
Annotation	White	Continuous	
Center	White	Center2	
Hatch	Cyan	Continuous	
Hidden	Blue	Hidden	
Object	Magenta	Continuous	
Viewport	Green	Continuous	No plot

5. Set Object layer current. Verify the Home tab is current. Choose the **Center - Diameter** option from the Circle command of the **Draw** panel as shown in Figure 9-12.

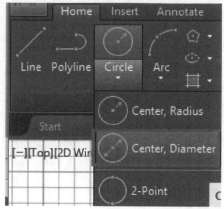

Figure 9-12

6. Draw the figure shown below on the Object layer. Draw a 4" diameter circle; place the center of the circle at 6,6. Repeat the Circle > Center > Diameter command and draw circles with 5" and 12" diameter with the center at 6,6 as shown in Figure 9-13.

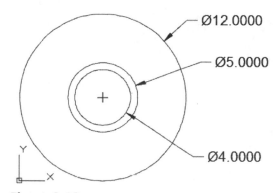

Figure 9-13

7. From the Status bar left click the flyout at p1 as shown in Figure 9-14 and toggle on Center and Quadrant object snap modes from the flyout menu.

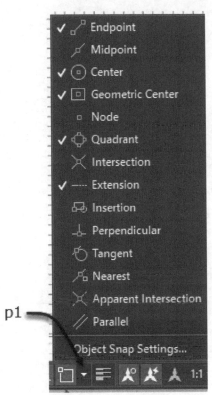

Figure 9-14

8. Use the Quadrant Object Snap mode to draw a line from the quadrant point shown in Figure 9-15 to the right 17 inches long.

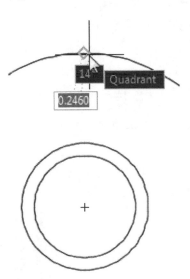

Figure 9-15

9. Continue to draw lines from the quadrant points to the right as shown in Figure 9-16 that are 17" long.

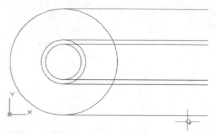

Figure 9-16

10. Toggle on the Endpoint object snap mode and draw the vertical line shown at p1 in Figure 9-17. Use the Offset command to copy the line at p1 to the left to create lines at p2 and p3 in Figure 9-17.

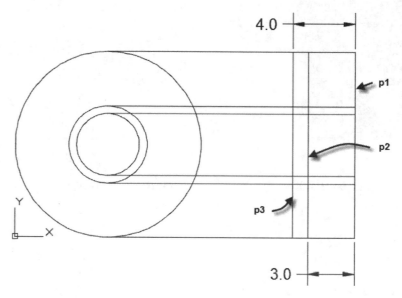

Figure 9-17

11. Use the Trim command to develop the right side view as shown in Figure 9-18. Do not dimension the geometry.

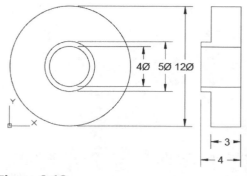

Figure 9-18

12. Verify Quick Properties is toggled on in the Status bar as shown in Figure 9-19.

Quick Properties

Figure 9-19

13. Choose the Centerline command from the Centerlines panel in the Annotate tab as shown at p1 in Figure 9-20. Respond to the workspace prompts as follows.
    Command: _centerline
    Select first line: (Select the line shown at p1 in Figure 9-20.)
    Select second line: (Select the line shown at p2 in Figure 9-20.)

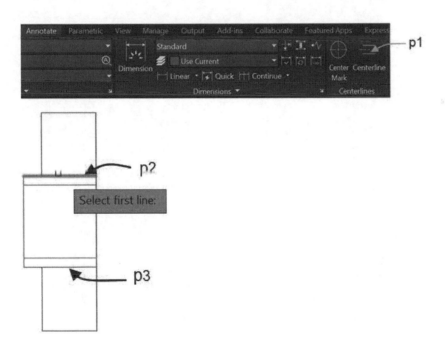

Figure 9-20

14. Select the line shown at p1, to display the Quick Properties palette as shown below. Edit the Start extension to 0.25 as shown at p2 and the End extension to 0.25 as shown at p3 in Figure 9-21. Edit the layer to Center as shown at p4 in Figure 9-21.

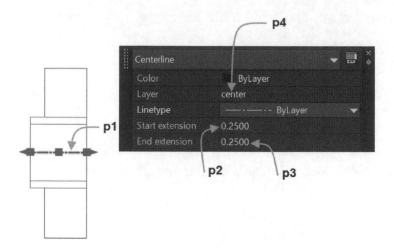

Figure 9-21

15. Choose Layout 1 from the lower left corner of the drawing window. Layout 1 is shown in Figure 9-22.

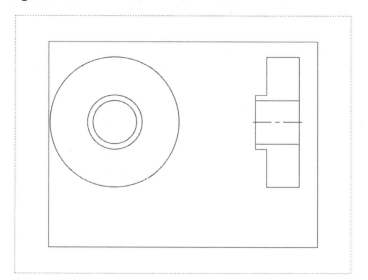

Figure 9-22

16. Right-click over the Layout 1 tab and choose **Rename**. Overtype **A 8.5 x 11** to name the layout as shown in Figure 9-23.

Figure 9-23

17. Right click over the A 8.5 x 11 layout; choose Page Setup Manager as shown in Figure 9-24.

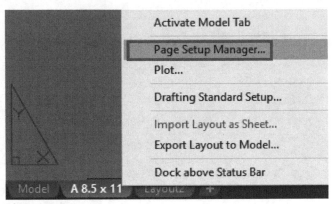

Figure 9-24

18. When the **Page Setup Manager** opens choose **Modify** shown in Figure 9-25 to specify settings for this layout.

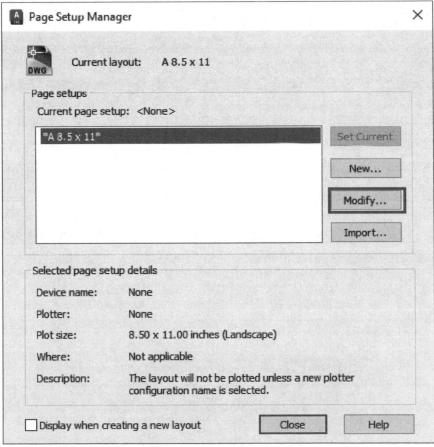

Figure 9-25

19. Edit the Page Setup to specify the printer to AutoCAD PDF (General Documentation).pc3 and set "What to plot" to **Layout** as shown in Figure 9-26. Click **OK** to dismiss the Page Setup dialog box. Click **Close** to close the Page Setup Manger.

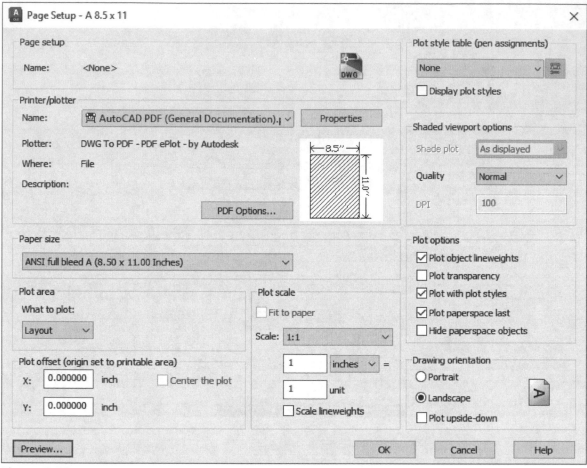

Figure 9-26

20. Select the viewport to display its grips. Choose the Viewport layer from the Layer flyout list of Quick Properties to change the layer to Viewport as shown in Figure 9-27. The viewport is displayed in green color.

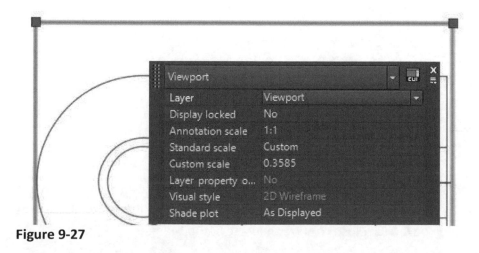

Figure 9-27

21. Press Escape to clear the display of grips and the selection of the viewport.

22. Verify the *Add scales to annotative objects when annotation scale changes* is toggled on as shown in Figure 9-28 in the Status bar.

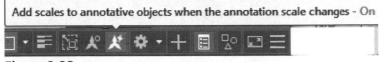

Figure 9-28

23. Double click inside the viewport. In this step you will change the scale of the viewport. Choose 1:5 from the Scale flyout as shown in Figure 9-29.

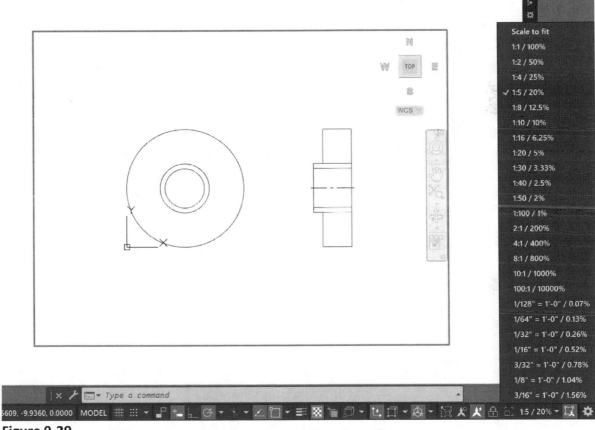

Figure 9-29

24. Note the center line of the pulley is displayed correctly as shown in Figure 9-30.

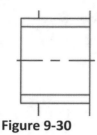

Figure 9-30

25. Right-click Layout 2, choose Rename. Overtype to name the layout Presentation.
26. Select the **Presentation** layout. Right click the Presentation and choose Page Setup Manager to open the Page Setup Manager and choose Modify shown in Figure 9-31 to specify the printer for the layout.

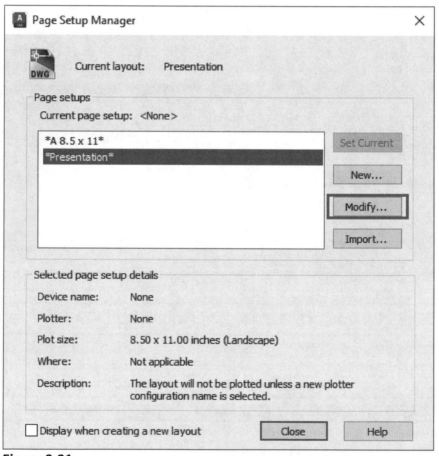

Figure 9-31

27. Edit the Printer/Plotter name to AutoCAD PDF (General Documentation).pc3 and Paper size to ANSI full bleed B (17.00 x 11.00 Inches) as shown in Figure 9-32. Verify the *What to plot* is set to Layout. Click OK to dismiss the Page Setup - Presentation dialog box. Click Close to dismiss the Page Setup Manager.

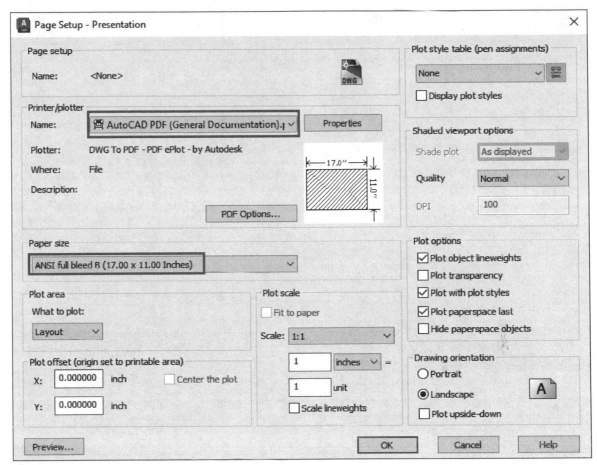

Figure 9-32

28. Double click outside the viewport to verify paper space is current.
29. Select the viewport. Select the grip at p1 as shown in Figure 9-33 and drag the grip to a location near p2.

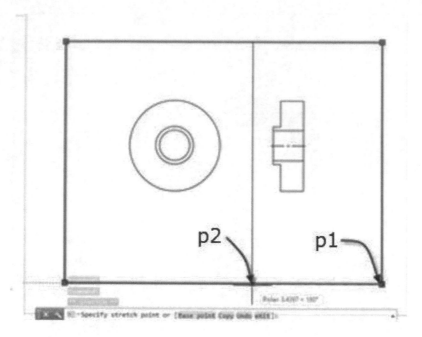

Figure 9-33

30. To change the layer assignment of the viewport, retain the selection of the viewport and choose Viewport name from the layer flyout list in the Quick Properties palette as shown in Figure 9-34.

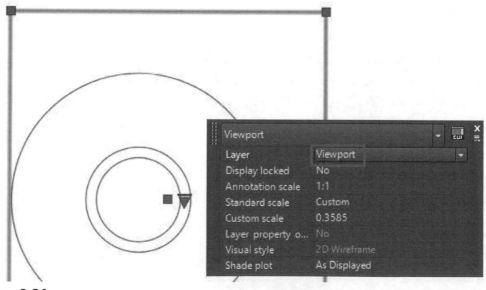

Figure 9-34

31. Double click inside the viewport; choose Zoom Extents from the Navigation bar to center the front and right-side views inside the viewport as shown in Figure 9-35.

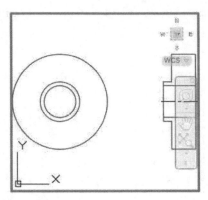

Figure 9-35

32. Choose the Model tab in the lower left corner of your screen.

33. Choose the Home tab. In this step you will add a centerline to locate the motor. Select the **Line** command and draw a line 4" from pulley shaft centerline as shown at p1 in Figure 9-36. Choose **Offset** from the Modify panel and respond to the workspace prompts as follows:

    Specify offset distance or [Through/Erase/Layer] <12>: 45 (Type 45 to specify the distance.)

    Specify point on side to offset or [Exit/Multiple/Undo] <Exit>: (Select a point above the object at p3 as shown in Figure 9-36.)

    Select object to offset or [Exit/Undo] <Exit>: *Cancel*(Press Enter to end the command.)

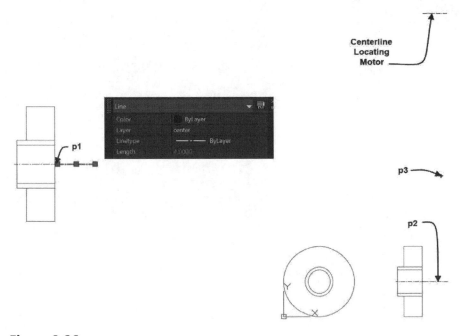

Figure 9-36

34. Select Presentation layout. Move the cursor to p1 as shown in Figure 9-37; double click to toggle to paper space.

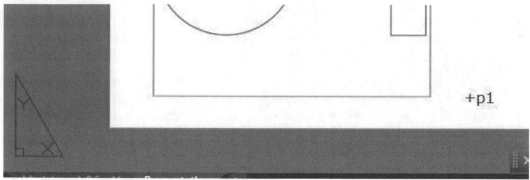

Figure 9-37

35. In the next series of steps, you will create a polygonal viewport.

36. Set the Viewport layer current. Choose the Layout tab from the Ribbon. Select the Polygonal tool from the Layout Viewports panel as shown at p1 in Figure 9-38.

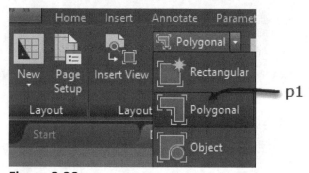

Figure 9-38

37. Sketch a closed polygon starting at p1 as shown in Figure 9-39.

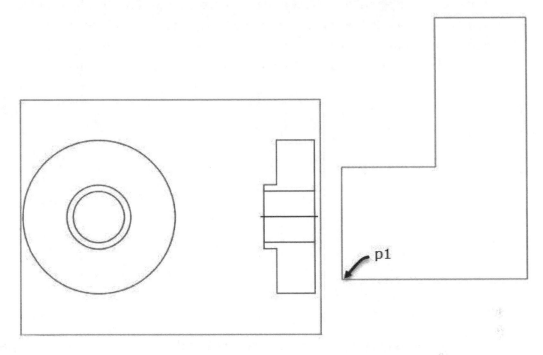

Figure 9-39

38. Upon completion of the viewport sketch a Zoom Extents is executed and entities are shown in Figure 9-40.

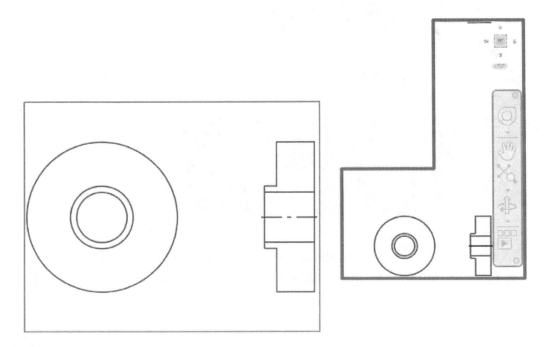

Figure 9-40

39. Double click inside the left viewport. Pan as necessary to view the object in the left viewport and choose the 1:5 scale from the Viewport Scale in the Status bar.

40. Double click inside the right viewport, choose the 1:10 scale from the Viewport Scale flyout and pan as necessary to view the centerline and part as shown in Figure 9-41.

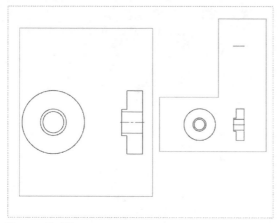

Figure 9-41

41. In the next series of steps, you will add dimensions that will only display in the viewport with the same scale as the annotation. In the Status bar verify that ***Show annotation objects – At current scale*** is set as shown in Figure 9-42.

Figure 9-42

42. Set the Annotative dimension style current in the Dimensions panel as shown in Figure 9-43.

Figure 9-43

43. Activate the left viewport and add a Diameter dimension as shown in Figure 9-44. Activate the right viewport and add a Linear dimension as shown in Figure 9-44.

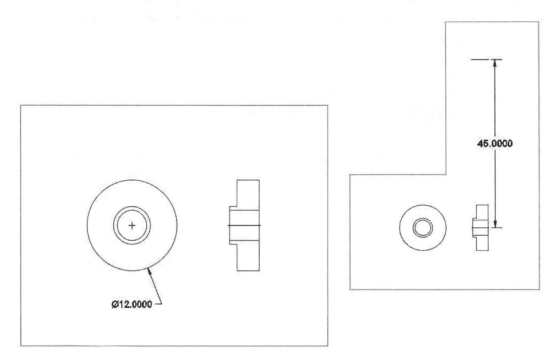

Figure 9-44

44. Note the dimensions per viewport are only displayed if the dimension scale representation matches the viewport scale. If you toggle the ***Show annotative objects to Always*** in the Status bar for a viewport all dimensions are displayed regardless of their scale representation as shown in Figure 9-45.

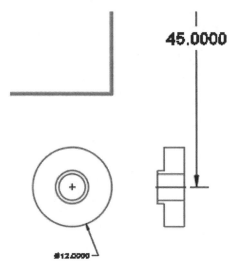

Figure 9-45

45. In the Status bar toggle off ***Show annotation objects – At current scale*** as shown in Figure 9-46.

Figure 9-46

46. Choose SaveAs from the Quick Access toolbar. Save the file as **Your Name Layout Viewport** in your student folder.

47. Close the file.

Plotting to Adobe PDF

The following tutorial provides an introduction to plotting and creating Adobe PDF files and the annotation scaling of text plotting content to scale.

1. Open AutoCAD 2024 from the Desktop shortcut.
2. Choose **Open** from the Quick Access toolbar.
3. Navigate to the **AutoCAD 2024 Certified User Exercise Files Dan Stine, Director of Design Technology, Lake|Flato Exercise Files \ Ch 9**; choose **Plotting.dwg**.
4. Set **Drafting & Annotation** workspace current in the Quick Access toolbar.
5. Choose Print > Plot from the Application Menu as shown in Figure 9-47.

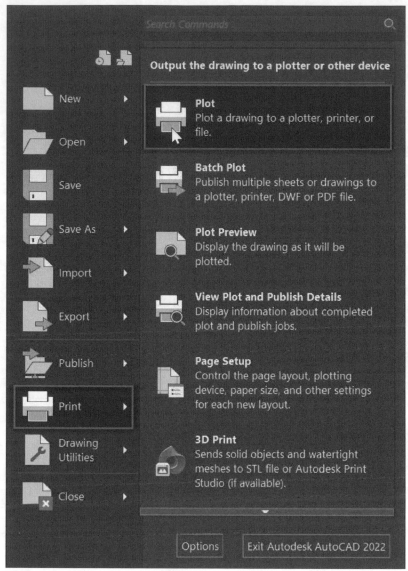

Figure 9-47

6. The Batch Plot message box opens as shown in Figure 9-48. Choose *Continue to plot as a single sheet as shown at p1* in Figure 9-48.

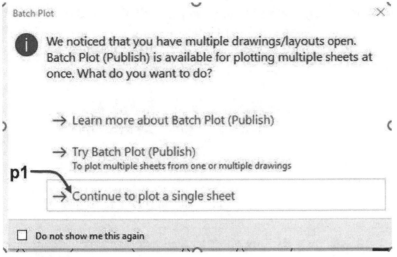

Figure 9-48

7. Edit the Plot Model dialog box as shown in Figure 9-49.

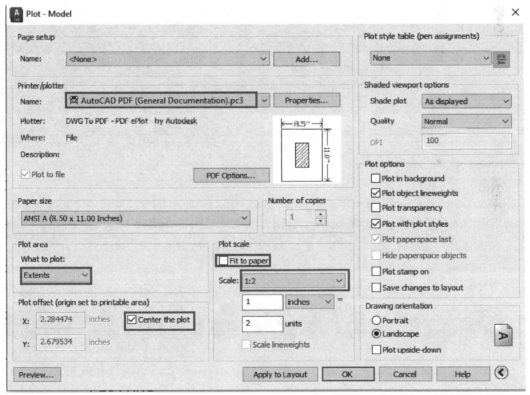

Figure 9-49

8. Choose **Preview** button of the Plot Model dialog box. Choose **Cancel** to dismiss the Plot – Plot Scale Confirm warning dialog box shown in Figure 9-50. Note, this warning is

displayed since the Annotation Scale of the drawing is set to 1:1 and the plot scale is set to 1:2. Choose Cancel to dismiss the Plot- Model dialog box.

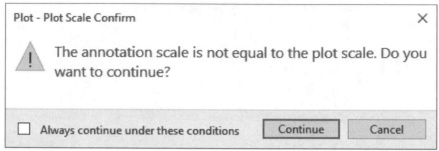

Figure 9-50

9. Select the Plotting and Scale text as shown at p1 in Figure 9-51 and edit Quick Properties palette as shown at p2 in Figure 9-51: Contents: PLOTTING \ PScale **1:2**, Annotative Style: choose **Annotative** from the flyout list, Annotative: select **Yes** from the flyout list. Press Escape to clear selection.

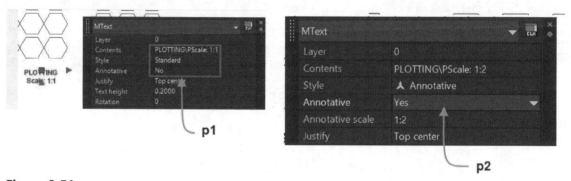

Figure 9-51

10. Edit the Add scales to annotative objects when the annotation scale changes ON and Annotation Scale to 1:2 in the Status bar as shown in Figure 9-52.

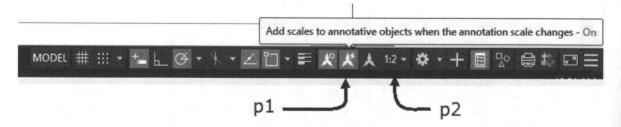

Figure 9-52

11. The text is scaled as specified in the Annotation Scale factor as shown in Figure 9-53.

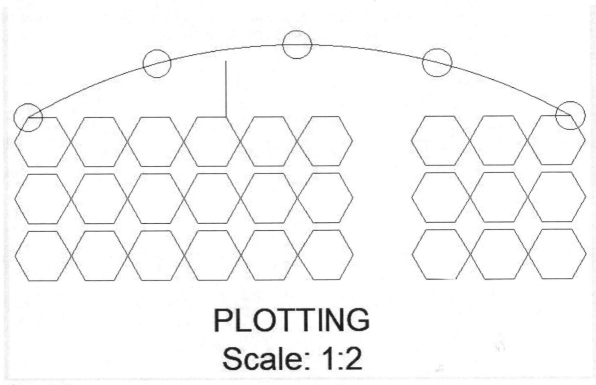

PLOTTING
Scale: 1:2

Figure 9-53

12. Choose Print from the Application Menu. The Batch Plot message box opens as shown in Figure 9-54. Choose Continue to plot as a single sheet.

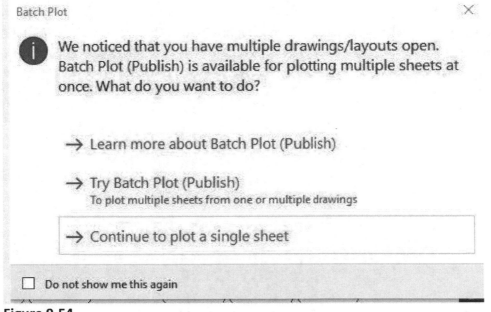

Figure 9-54

13. Edit the Plot Model dialog box as shown in Figure 9-55. Choose the Preview button located in the lower left corner of the Plot Model dialog box.

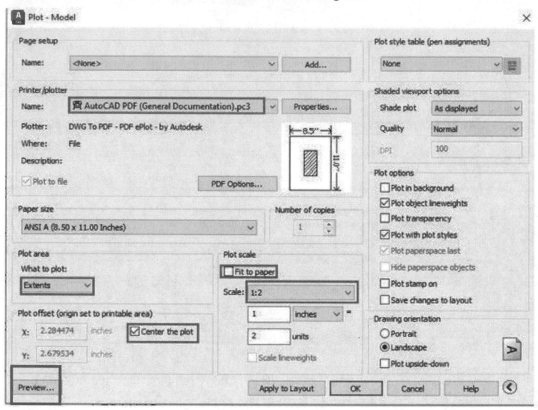

Figure 9-55

14. The preview of the drawing is shown in Figure 9-56. If the preview image matches the image of Figure 9-56 choose the Close window button of the toolbar located at the top left corner as shown at p1.

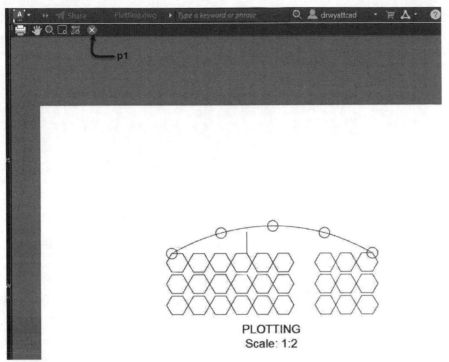

PLOTTING
Scale: 1:2

Figure 9-56

15. The Plot Model dialog box reopens; choose OK to dismiss the dialog box. The Browse for Plot File dialog box opens as shown in Figure 9-57. Navigate to your Student folder and edit the name to include your name in front of the plot file name as shown in Figure 9-57.

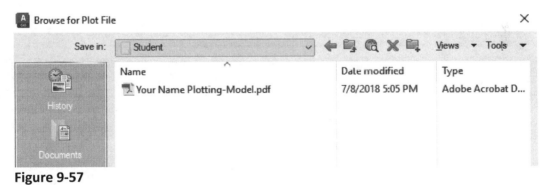

Figure 9-57

16. The Your Name Plotting- Model.pdf file opens in preview as shown in Figure 9-58.

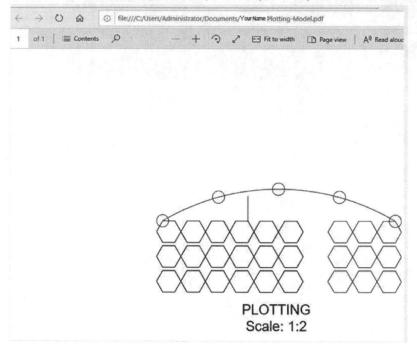

Figure 9-58

Chapter 10
Practice Exam

This chapter will describe the practice exam provided with the text including:

- How to access the practice test
- Downloading of files needed for the exam
- The test screen layout
- Grading and feedback from a test attempt

Practice Test Content

The practice test includes multiple-choice questions, hot spots of the AutoCAD menu, and fill-in-the-blank questions. The fill-in-the-blank questions require you to determine distances, coordinates, or areas after you perform edits to entities using AutoCAD and sample drawing files.

Accessing the Practice Test

Prior to downloading the test, you should create a folder on your computer. For the example below, I created a folder at C: Documents\ ACU \ AutoCAD \YOUR NAME_Student \.

Download the test from the publisher's website as per the instructions found on the inside front cover of this book, using the unique access code provided.

The test and drawing files required for the test are located in a zipped file, which includes the folder for the test content. Select the zip file, right-click, and choose **Extract All** from the contextual menu, as shown at p1 in Figure 10-1.

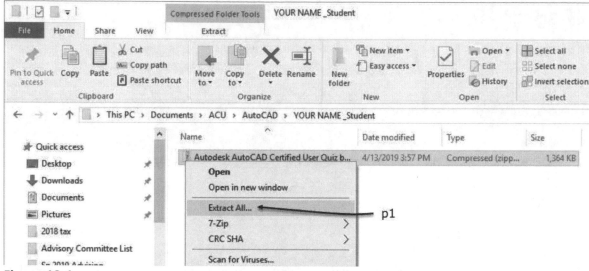

Figure 10-1

The Select a Destination and Extract Files dialog box opens, prompting you to select the location of the folder for placement of the test files and support files. Choose the Browse button shown in Figure 10-2 to verify or change to your student folder C: Documents\ ACU \ AutoCAD \YOUR NAME_Student\. When you choose the Extract button, the folder of material is copied to the C: Documents\ ACU \ AutoCAD \YOUR NAME_Student\. The default location is to place the extracted folder in the same folder as your zip file. You can choose the Browse button to change the location. If the location is acceptable, choose the Extract button shown at p1 in Figure 10-2.

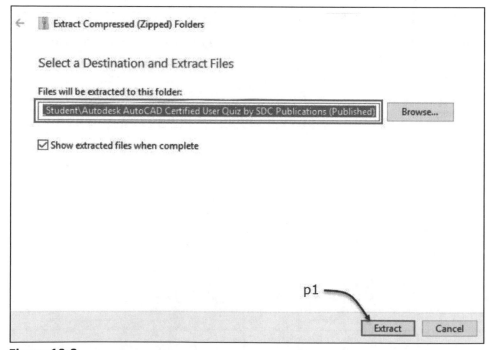

Figure 10-2

Double-click on the Autodesk AutoCAD Certified User Quiz by SDC Publications (Published) folder to open it. Note that the duplicate folder name is created when the folder is zipped. The content extracted will include an AutoCAD Practice Test Files folder, a data folder, and a file named **index.html**, as shown in Figure 10-3. The index.html file will open the test with Windows Explorer or Google Chrome (on a PC or Mac).

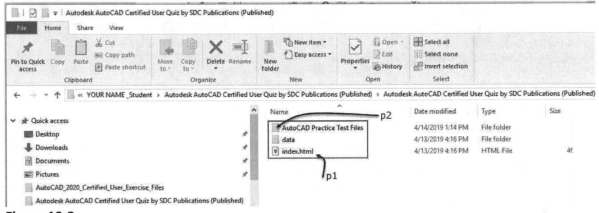

Figure 10-3

Some questions on the test can be answered without the AutoCAD software. The test includes sample questions from each of the objective categories, which are shown in Figure 10-4. If you have dual monitors, place the test on one monitor and AutoCAD on the other. If you have a single monitor, you can toggle to different applications by pressing the **Alt + Tab** keys.

AutoCAD Exam Objectives

Draw and Modify Objects
- Create basic drawing objects
- Draw polylines
- Select and deselect objects
- Manage layers
- Work with blocks

Draw with Accuracy
- Apply basic object snaps
- Identify and use coordinates

Basic Editing
- Modify object properties
- Use basic editing commands to modify objects
- Trim, extend, or lengthen objects
- Create rectangular and polar arrays
- Offset objects at a specific distance
- Apply a fillet or chamfer to objects

Annotation
- Create and modify text
- Add and modify leaders and/or multileaders
- Create and edit dimensions
- Apply hatches or fill patterns

Layouts and Printing
- Work with layouts and viewports
- Manage output formats

Figure 10-4

Accessing AutoCAD Files

When you are ready to start the exam you should open AutoCAD, choose Open from the Quick Access toolbar and navigate to the C: Documents\ ACU \ AutoCAD \YOUR NAME_Student \ Autodesk AutoCAD Certified User Quiz by SDC Publications (Published)\ Autodesk AutoCAD Certified User Quiz by SDC Publications (Published) \ AutoCAD Practice Test Files as shown in Figure 10-5. The following files located in the AutoCAD Practice Test Files folder are required for the test: ABC Template.dwt, Test 1.dwg, Test 2.dwg and Test 3.dwg.

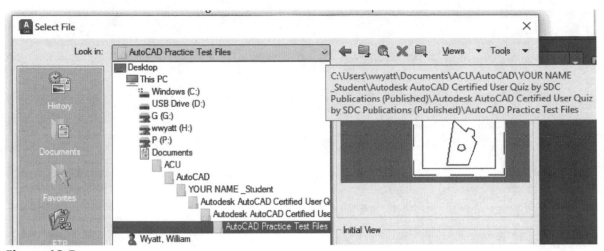

Figure 10-5

Select each of the AutoCAD drawing files and choose **Open**. The drawing tabs should be displayed below the ribbon for each open file as shown in Figure 10-6.

Figure 10-6

Since the path to AutoCAD Practice Test Files is very long, I suggest you use the Places panel to specify and retain the location. To open the template file, choose Open from the Quick Access Toolbar. When the Select File dialog box opens, verify that the AutoCAD Practice Test Files is current in the Look in field. Right-click over the Places panel as shown at p1 in Figure 10-7 and choose Add Current Folder as shown in the contextual menu in Figure 10-7.

Figure 10-7

Edit the Files of type drop-down to select Drawing Template (.dwt) shown at p1 in Figure 10-8. The files listed immediately switch to the AutoCAD template folder. Choose the AutoCAD Practice Test Files button shown at p2 of the Places panel in Figure 10-8.

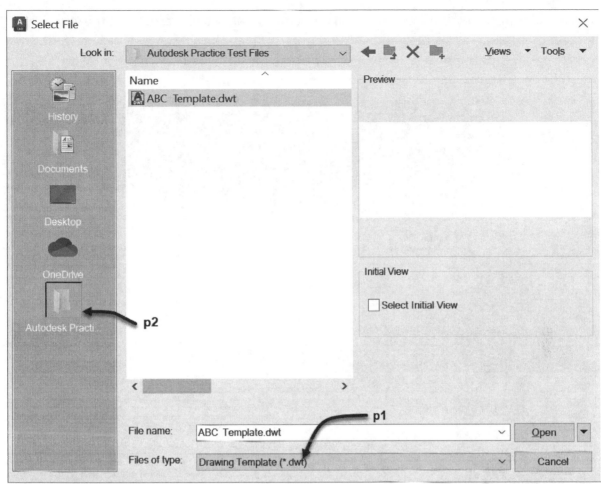

Figure 10-8

Verify that all four drawing file tabs are displayed as shown in Figure 10-9.

Figure 10-9

Each question, which requires the use of AutoCAD, will include the drawing name and layout name, as shown in Figure 10-10. The layout will include a viewport, which is a fixed location for you to edit the AutoCAD drawing file for a question. The view of the content of the viewport is locked; therefore, you should not change the zoom or pan. The question will require you to perform AutoCAD commands and upon completion, to enter distances, areas or object information using the Inquiry commands or the Properties palette.

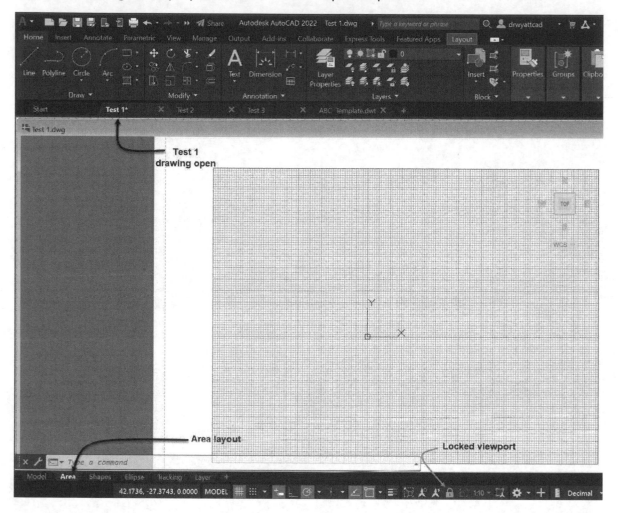

Figure 10-10

Starting the Exam

To start the exam, you must be connected to the internet and have the internet browser opened. Open the C: Documents\ ACU \ AutoCAD \YOUR NAME_Student \ Autodesk AutoCAD Certified User Quiz by SDC Publications (Published)\ Autodesk AutoCAD Certified User Quiz by SDC Publications (Published) and right-click the index.html file. Choose Open with > Google Chrome to open the test.

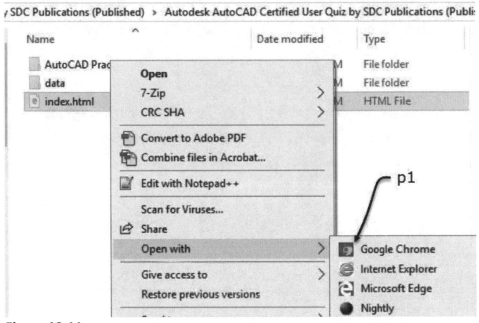

Figure 10-11

When you open the file, the test will open as shown in Figure 10-12 in your internet browser. Click the **Start Quiz** button located at p1 as shown in Figure 10-12.

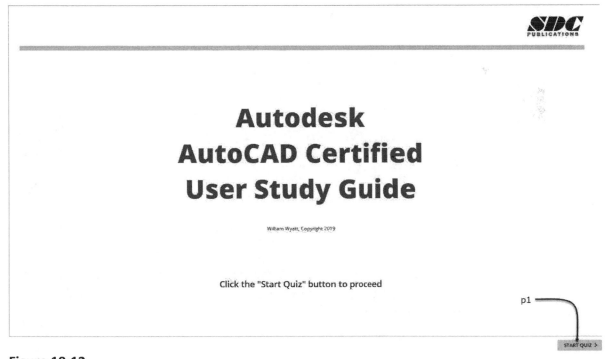

Figure 10-12

Enter your name and email address at p1 in the Details page, as shown in Figure 10-13, if you would like to send your results to your teacher or to retain a copy. Click the Submit button as shown at p2 in Figure 10-13.

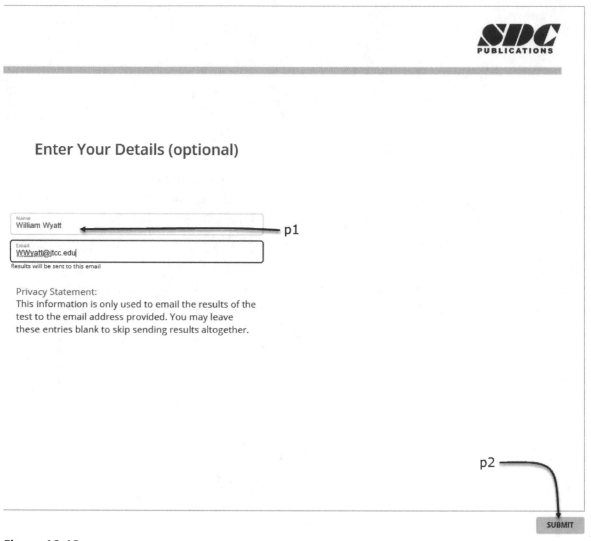

Figure 10-13

The following tips are included on the Quiz Instructions page. Carefully read the questions and then answer each question. Questions that require you to enter distances or areas will be marked as correct only if you match the required format, as shown in Figure 10-14. Click the Continue button shown in the lower right corner to start the first question.

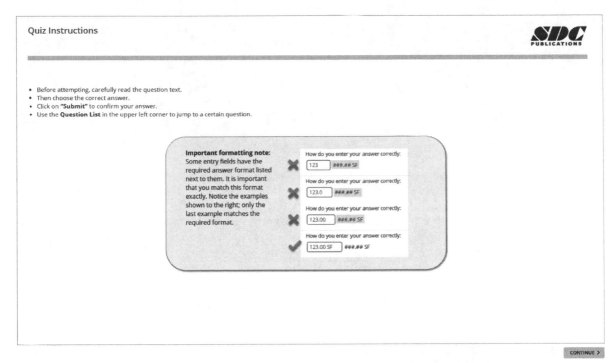

Figure 10-14

As you respond to each question, you will be notified immediately if you answered correctly. If you answered the question incorrectly, you will be directed to refer to the chapter and pages of the book in which the topic was presented.

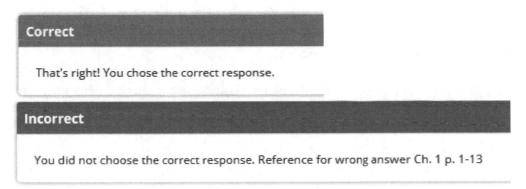

Figure 10-15

You can review your success for all questions by choosing the Questions icon located in the upper left corner of each question, as shown at p1 in Figure 10-16. The question list, the points awarded, and the results are shown at p2 in Figure 10-16.

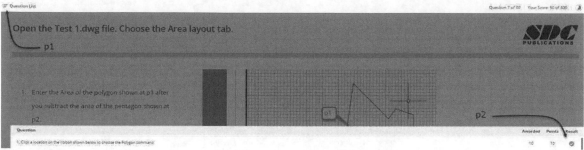

Figure 10-16

The current question number is shown at the top right header of each page as shown in Figure 10-17, along with your current score and time remaining for the test session. You must respond to each question to move on to the next question. You cannot back up to previous questions during the test.

Figure 10-17

Using the Results for Additional Preparation

Upon completion of the test you will be immediately notified if you passed or failed as shown in Figure 10-18. Your score will be listed, and the acceptable passing score as shown in Figure 10-18. You must answer 80% of the questions correctly in order to pass the quiz; however, the ACU exam requires 70% correct. You can choose the Review Quiz button shown in Figure 10-18 to view the question list as shown at p2 in Figure 10-19 to review each question. The results of each question are shown at the bottom of Figure 10-19 at p1 for the questions highlighted in Figure 10-19.

You did not pass.

Your Score: **17% (50 points)**

Passing Score: **80% (240 points)**

REVIEW QUIZ

Figure 10-18

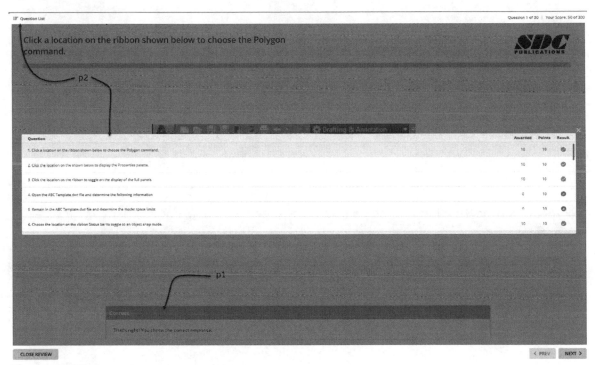

Click a location on the ribbon shown below to choose the Polygon command.

Figure 10-19

Your results will be emailed to you if you entered an email address before taking the exam, which includes the score, time spent, and the list of questions marked as correct or incorrect. You can use the emailed results to identify weak areas for continued study.

Here you can review the quiz results for "Autodesk AutoCAD Certified User Quiz by SDC Publications".

Name   William Wyatt
Email   wwyatt@jtcc.edu

Date/Time: November 30, 2021 9:50 PM
Answered: 26 / 30
Your Score: 260 / 300 (87%)
Passing Score: 240 (80%)
Time Spent: 45 min
Result   Passed

Question 1 Correct
Points: 10/10
Click a location on the ribbon shown below to choose the Polygon command.

Your Answer	Correct Answer
Polygon Command	Polygon Command

Question 2 Correct
Points: 10/10
Click the location on the shown below to display the Properties palette.

Figure 10-20

Remember, certification benefits students as they seek their initial employment and future promotions.